EXCITONS

Edited by **Sergei L. Pyshkin**

Excitons
http://dx.doi.org/10.5772/66232
Edited by Sergei L. Pyshkin

Contributors

Shota Ono, Kaoru Ohno, Peter Wachter, Thibaud Etienne, Sabyasachi Kar, Yew Kam Ho, Sergei L. Pyshkin

Published by InTech
Janeza Trdine 9, 51000 Rijeka, Croatia

Publishing Process Manager Nina Kalinic
Technical Editor SPi Global
Cover InTech Design team

First published March, 2018
Printed in Croatia
Legal deposit, Croatia: National and University Library in Zagreb

Additional hard copies can be obtained from orders@intechopen.com

Excitons, Edited by Sergei L. Pyshkin
p. cm.
Print ISBN 978-953-51-3869-3
Online ISBN 978-953-51-3870-9

For EU product safety concerns:
IN TECH d.o.o., Prolaz Marije Krucifikse Kozulić 3, 51000 Rijeka, Croatia,
info@intechopen.com or visit our website at intechopen.com.

Meet the editor

Dr. Sc. Prof. Sergei L. Pyshkin is a scientific advisor at the Institute of Applied Physics, Academy of Sciences of Moldova. He is an adjunct professor and senior fellow at the Clemson University, SC, USA. He is a member of the US Minerals, Metals & Materials Society (TMS). He received the State Prize of Republic of Moldova Award for investigations in solid-state physics and microelectronics. The works deal with nonlinear optics (multiquantum absorption), electron and phonon transport phenomena, photoconductivity and light scattering, luminescence, crystal and thin film growth, molecular beam and laser-assisted epitaxies, nanotechnology, lasers for medicine, and scientific instrument making (boxcar integrators and solid-state IR matrix photoreceivers). Biography of Prof. Pyshkin is included in the Marquis "Who's Who in America" (2008–2013) and "Who's Who in the World" (2009–Present).

Contents

Preface

"Excitons," InTech Open Access book, consists of exciting complementary perspectives on the progress in the field of excitons and their use in processes occurring in modern optoelectronic device structures, with contributions from authors from France, Switzerland, Moldova, China, and Japan.

Emeritus Prof. Wachter Peter, Switzerland, investigates and explains the heat conductivity, thermal diffusivity, compressibility, sound velocity, and exciton-polaron dispersion, while Dr. Etienne Thibaud, France, develops the problem of qualitative and quantitative topological analyses of molecular excitons. Dr. Ono Shota and Kaoru Ono, Japan, in their turn, elaborate on exciton dissociation in organic solar cells, while scientists from the Republic of China (Taiwan), Dr. Sabyasachi Kar and Yew Kam, consider the interesting analogy of exciton, biexciton, and trion to the positronium atom, molecule, and negative ion. The advantages and recent progress in these areas, which are important and exciting problems currently under investigation in the field of excitons, are convincingly presented and discussed.

The potential for the use of excitons for the future technology is hard to underestimate. The introductory chapter "Bound Excitons in Gallium Phosphide" presented by the editor of the book, Prof. Sergei L. Pyshkin, scientific advisor at the Institute of Applied Physics, Academy of Sciences of Moldova, and adjunct professor of Clemson University, SC, USA, discusses, among other issues, that both study and application of the exciton properties are a difficult task, mainly due to the low quality of freshly prepared semiconductor crystals. Freshly prepared crystals are usually characterized with a large concentration of crystal structure defects, such as vacancies and dislocations of the proper arrangement of intrinsic and impurity atoms in the artificially grown crystal structures. Thus, despite all efforts of crystal growth experts, it is virtually impossible to compete with natural crystals grown for thousands of years in favorable natural conditions. The results of over 50 years of investigations of a unique set of GaP semiconductor samples are also presented in the book. The discussion highlights the significant improvement in the properties of GaP:N crystals prepared in the 1960s through the formation of the perfect host crystal lattice and the N-impurity crystal superlattice or of an excitonic crystal. A new approach to the selection and preparation of perfect materials for optoelectronics is described, offering a unique opportunity of a new form of solid-state host—the excitonic crystal—as a high-intensity light source with low thresholds for nonlinear optical effects.

Generally, the book highlights the fact that excitonic crystals yield novel and useful properties including enhanced stimulated emission and very bright and broadband luminescence at room temperature. The discussion presented in the book is inspired by many outstanding scientists, including the prominent in the field of exciton topics late Prof. Leonid V. Keldysh, and other well-known colleagues, representing numerous scientific centers worldwide, particularly Russia, the USA, and Italy, who at various points made invaluable contribution to understanding and advancing the ideas in the field of excitons. The studies presented are relevant due to the unprecedented interest of researchers from all over the world in using excitons and their properties in optoelectronics, nanoscience, and technology, such as in the development of modern optoelectronic device structures. They are also relevant and interesting for the representatives of both public and private sectors as they offer a significant contribution to high-technology driven industries.

Dr. Sc. Prof. Sergei L. Pyshkin
Adjunct Professor and Senior Fellow
Clemson University, SC, USA

Scientific Advisor at the Institute of Applied Physics
Academy of Sciences of Moldova, Moldova

Chapter 1

Introductory Chapter: Bound Excitons in Gallium Phosphide

Sergei L. Pyshkin

Additional information is available at the end of the chapter

http://dx.doi.org/10.5772/intechopen.73550

1. Introduction

The authors contributing to InTech open access book *Excitons* offer exciting complementary perspectives on the progress in the field of excitons and their use in processes occurring in modern optoelectronic device structures. This is both an important and a complex field, as will be elaborated further on, which is why it has been chosen as an introductory stance to summarize some findings in the field made by the author of this chapter, also the editor of this particular book.

As we note the unprecedented interest of researchers from all over the world in using excitons in the development of modern optoelectronic device structures, we offer some of the results and material gathered in the process of our half-a-century long work for further study and application in electronic companies. The results presented here and in References to this Chapter are inspired by many outstanding scientists, my teachers and the colleagues, representing a number of scientific centers worldwide and in particular Russia, the USA, and Italy, who at various points made invaluable contributions to understanding and advancing the ideas on results obtained through the years of my research.

We have been growing and exploring gallium phosphide [1–8] for more than a half a century, a process of experimenting, analysis and observation which resulted in unique material reflecting previously unexplored properties of excitons and new prospects for the use of GaP, which could be very interesting for application in the electronic industry.

Studying and using new properties of excitons are a difficult task, mainly due to the low quality of freshly prepared semiconductor and other crystals. Fresh crystals are usually characterized with a large concentration of crystal structure defects, such as vacancies and dislocations of the proper arrangement of intrinsic and impurity atoms in the GaP (face-centered cube) crystal structure. Notably, despite all efforts of crystal growth experts, it is virtually impossible to compete with natural crystals grown for thousands of years in favorable natural conditions. Large concentration of defects, noted above, inevitably arises from the rapid freezing of the constituent crystals of the atoms and the dopants in positions, which they occupy being in the liquid GaP phase at the time of the beginning of cooling and the formation of embryos, according to the adopted technology for obtaining crystals of GaP [1].

Natural tendency of own and impurity atoms to occupy the places assigned to them by the crystal lattice is hampered by their infinitesimally slow diffusion rate at room or at low temperature of storage of grown crystals. In this way, decades pass before the lattice component occupies the exactly correct position in the crystal lattice, diffusing inside it at the storage temperature, from the place where it was at the time of the onset of cooling of a mixture of GaP, necessary for the onset of deposition and further growth of pure or doped crystals (see details in [1–8]). Naturally, most crystal manufacturers are reluctant to wait for improvements in the structure and properties of imperfect crystals, as this process is extremely slow. This fact leads to the need of using noncompliant materials with poor parameters, which however drastically reduces the lifetime and quality of device structures made from them, and in addition increases production costs and sharply reduces the value of the output into the electronic industry. For instance, presently manufactured low-quality materials cause high margin of error in microchips resulting in quality problems with microelectronic-based devices, such as mobile phones and computers, but also devices used in "heavy" industries, such as healthcare, space, or defense. Due to the described limitations, the possibility of using excitons as the most vulnerable material easily destroyed by defects of photon carriers is significantly reduced.

Taking the aforementioned into account, we consider some properties of bound excitons in GaP, including the possibility and the expected results of their application in optoelectronic device structures. Recall that the term *bound exciton* means an electron-hole pair localized near the impurity center. In our case this is an isoelectronic impurity N replacing the own P atom in the GaP lattice, possessing a giant-capture cross section with respect to the free electron. The captured electron attracts a hole, forming a bound exciton.

The presence of a heavy nucleus (atom N^{+} trapped electron) is an important feature of bound excitons, which, under appropriate conditions, allows them to form a solid exciton phase, in contrast to free excitons, where the transition to the solid phase is impossible due to the approximate equality of the effective masses of the electron and holes and so-called zero-point oscillations, which destroy our attempts to form a solid phase with further condensation of a system of coupled excitons. We also note the possibility of creating exciton crystals that arise in the ordered arrangement of impurity centers and the creation of an appropriate impurity sublattice with a crystal structure analogous to the GaP single crystal, but with a lattice parameter equal to the Bohr diameter of the bound exciton in this material (approximately 10 nm).

Keeping in mind possibly groundbreaking features (at least for some industries, we have already mentioned) of the solid exciton phase, we have focused our long-term technological efforts on obtaining perfect GaP crystals, including creation and investigation of the properties of perfect GaP crystals and certain device structures based on them. We have established that the mentioned above impurity sublattice arises with prolonged storage of GaP single crystal plates under natural conditions (room temperature and normal pressure). According to our estimates [2, 6], the crystals must be 10–15 years old under these particular conditions. During this time, randomly distributed impurities form the correct crystalline sublattice at room temperature, due to the natural diffusion of impurities into places with their low concentration and their displacement to places that reduce mechanical tensions in the crystal. Despite the fact that this process takes over a decade, the qualities of the output material offer numerous opportunities in some of the key industries based on microelectronic technologies, especially strategic, long-cycle ones. Even though current expectations of the cycle of perfect crystal growth may last up to 15 years or more, we strongly believe that this long-term process can be significantly shortened by skillful selection of storage conditions (accelerating the diffusion temperature, applying counterpressure using the vapor of volatile components P, etc.). Masterful selection of storage conditions and the eventual drastic reduction in the time needed for obtaining close to ideal crystals, along with other factors considered below, incentivize to test and potentially introduce the proposed method of nearly perfect crystal growth into some key electronic industries and make devices of highest quality based on the top quality crystals.

In addition to the above, after 10–15 years from the beginning of the introduction of the proposed system for obtaining perfect crystals, there will be no need to wait for decades to "mature" them. This is possible provided that new materials are stored permanently and only materials that have been ripening for a certain period of time, which according to our estimates are 10–15 years old, are taken out for use.

In order to incentivize further interest, part of the methodology is described in short as follows. We have used a sublattice of N atoms at distance of 10 nm prepared in advance in the crystal and engaged powerful optical pumping with photon energy exceeding the width of the forbidden band of GaP and the power of light sufficient to fill all the impurity centers. In this way, we were able, for the first time in global practice, to obtain an excitonic crystal, schematically depicted in **Figure 1**. In addition, **Figure 1a** shows phosphorus host atoms P in the GaP crystal lattice and atoms of nitrogen impurity N periodically replacing them through 10 nm. The electrons trapped by the impurity centers and the holes interacting with them at corresponding excitation level form an excitonic crystal shown in **Figure 1b**.

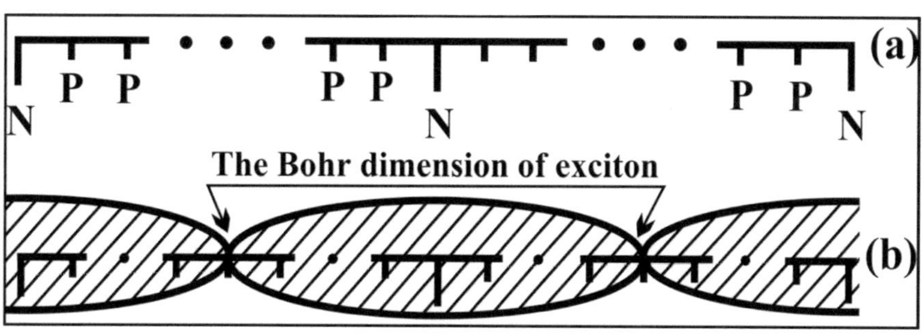

Figure 1. Models of the well-ordered GaP:N [4, 8]. (a) The new type of crystal lattice with periodic substitution of N atoms for the host P atoms. (b) The excitonic crystal on the basis of this superlattice. Substitution period is equal to the Bohr diameter of exciton (~100 Å), and optical excitation is enough for complete saturation of the N sublattice with nonequilibrium electron-hole pairs (see details in Ref. [8]).

Note that none of the nanotechnology methods are used in the creation or selection of dimensions of these nanoparticles but only natural forces of electron-hole interaction. As the result, we get something like a neutral short-lived crystal nuclei (N atoms with captured electrons) and holes, interacting with them through Coulomb forces. The so-called zero vibrations do not destroy this solid phase of bound excitons having these heavy nuclei that give an opportunity to reach their crystal state—short-lived excitonic crystal.

It is interesting to compare the luminescence of freshly prepared, partially (a) and ideally ordered GaP:N (b) crystals presented in **Figure 2**.

We note that the same freshly prepared crystals do not possess any luminescence at all because of the enormous number of defects that supply the radiationless return to the valence band of electrons excited by light. The same partially ordered crystals exhibit a luminescence spectrum of excitons consisting of a zero-phonon line and its phonon satellites in the emission of the intrinsic acoustic and optical phonons of the GaP lattice (**Figure 2a**).

Earlier, we observed a clear stimulated emission from a GaP:N resonator at 80 K [4] in freshly prepared crystals, as well as the so-called superluminescence from the GaP single crystals. Presently, our ordered crystals have a bright superluminescence at room temperature that implies their perfection and very **lower** light losses. Thus, we demonstrate that stimulated emission is developed even at room temperature by direct electron-hole recombination of an electron at the bottom of the conduction band with a hole at the top of the valence band and the LO phonon absorption.

We also note an interesting analogy between the radiation of a well-ordered and perfect GaP crystal and well-prepared nanoparticles based on it, which we present in **Figure 3**.

Thus, sticking to some specific rules, including the necessity to build single crystal the excitonic superlattice with the identity period equal to the bound exciton Bohr dimension in the GaP:N, we get a unique opportunity to create a new solid state media consisting of short-lived nanoparticle excitonic crystal. It, obviously, has very useful and interesting properties for application in optoelectronics, nanoscience, and technology.

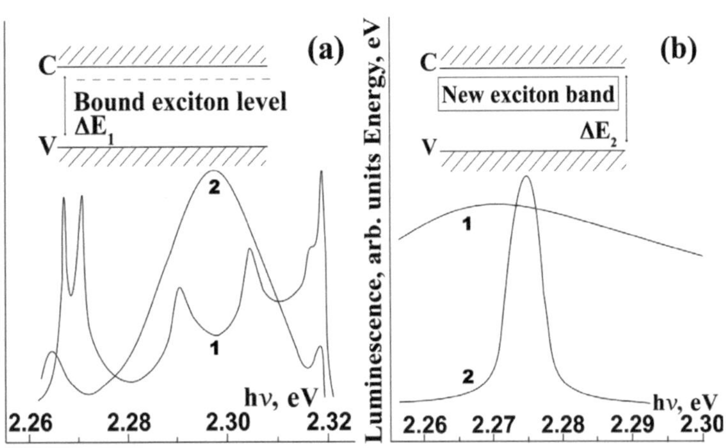

Figure 2. Luminescent spectra and schematic representation of the forbidden gaps (ΔE1, ΔE2) in the nitrogen-doped GaP aged for (a) 25 years and (b) 40 years.

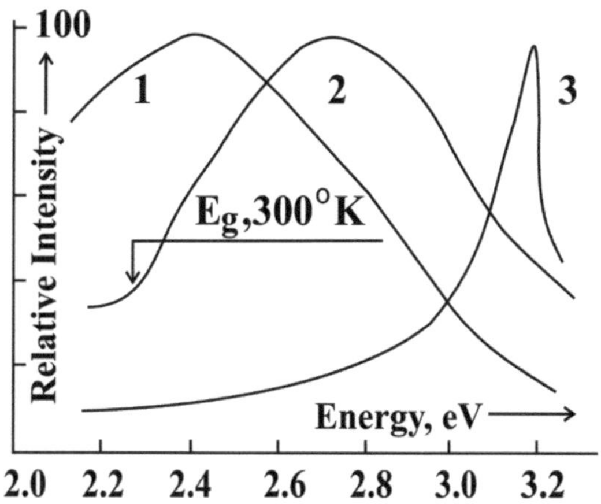

Figure 3. Luminescence of perfect bulk GaP single crystals [1] in comparison with the luminescence of GaP nanoparticles and GaP/polymer nanocomposites [2, 3]. Prepared by us, nanoparticles [5, 7] were stored as dry powder (spectrum 2) or suspension in a liquid (spectrum 3).

Acknowledgements

We are happy to note that the broad discussion and dissemination of our joint results stimulate further collaboration with our partners from the USA, Russia, Italy, Romania, France, and other countries.

Prof. Sergei L. Pyshkin expresses his cordial gratitude to his teachers, renowned scientists late Prof. Nina A. Goryunova, Nobel Prize Laureate Alexander M. Prokhorov, and academicians Leonid V. Keldysh, Rem V. Khokhlov, and Sergei I. Radautsan, as well as to the US Department of State; Institute of International Exchange, Washington, DC; the US Air Force Office for Scientific Research; the US Office of Naval Research Global; Civilian R&D Foundation, Arlington, VA; the US Science & Technology Center in Ukraine; his colleagues and coauthors from Clemson University, SC; University of Central Florida, FL; Istituto di elettronica dello stato solido, CNR, Rome, Italy; Universita degli studi, Cagliari, Italy; Lomonosov Moscow State University; Joffe Physico-Technical Institute and State Polytechnical University, St. Petersburg; Ac. Scie. Institute of General Physics, Moscow, Russia; and Institute of Applied Physics and Academy of Sciences of Moldova for support and attention to this protracted (1961 to present time) research.

Author details

Sergei L. Pyshkin[1,2]*

*Address all correspondence to: spyshkin@yahoo.com

1 Institute of Applied Physics, Academy of Sciences, Kishinev, Moldova

2 Clemson University, South Carolina, United States of America

References

[1] Goryunova NA, Pyshkin SL, Borshchevskii AS, et al. Influence of impurities and crystallisation conditions on growth of platelet GaP crystals. Growth of Crystals. 1969;**8**:68-72. In: Sheftal NN, editor. Symposium on Crystal Growth at the Seventh Int Crystallography Congress; July 1966; Moscow. New York: Consultants Bureau. Rost crystallov, 1968;**8**(2):84 (in Russian)

[2] Pyshkin SL, Radautsan SI, Zenchenko VP. Processes of long-lasting ordering in crystals with a partly inverse spinel structure. Soviet Phys – Doklady. 1990;**35**(4):301-304

[3] Pyshkin SL. Preparation and properties of gallium phosphide. [Ph.D. thesis]. Kishinev: State University of Moldova; 1967

[4] Pyshkin SL. Photoconductivity and luminescence of highly optically excited semiconductors. [Dr.Sc. thesis]. Lomonosov Moscow State University; 1978

[5] Pyshkin SL, Ballato J. Long-term convergence of bulk- and nano-crystal properties. Chapter 19. In: Optoelectronics – Materials and Technics. Rijeka: InTech – Open access Publisher; 2011. pp. 459-476. ISBN: 978-953-307-276-0

[6] Pyshkin S, Ballato J. Advanced light emissive device structures. Chapter 1. In: Pyshkin SL, Ballato J, editors. Optoelectronics – Advanced Materials and Devices. Rijeka: InTech – Open Access Publisher; 2013. pp. 1-24. ISBN: 978-953-51-0922-8

[7] Pyshkin SL. Excitonic crystal and perfect semiconductors for optoelectronics. Chapter 1. In: Pyshkin SL, Ballato J, editors. Optoelectronics – Advanced Materials and Devices. Rijeka: InTech – Open Access Publisher; 2015. pp. 1-30. ISBN: 978-953-51-0922-8

[8] Pyshkin SL, Ballato J. Properties of GaP studied over 50 years. Chapter 1. In: Pyshkin SL, Ballato J, editors. Optoelectronics – Advanced Device Structures. Materials and Devices. Rijeka: InTech – Open Access Publisher; 2017. pp. 1-20. ISBN: 978-953-51-3369-8

Chapter 2

Exciton Condensation and Superfluidity in $TmSe_{0.45}Te_{0.55}$

Peter Wachter

Additional information is available at the end of the chapter

http://dx.doi.org/10.5772/intechopen.70095

Abstract

In this publication, details of the calculation of heat conductivity and thermal diffusivity, compressibility, sound velocity and exciton-polaron dispersion will be shown. The properties of excitons, coupling to phonons, producing thus polarons, but also blocking the phonons as running waves lead to an exciton condensation or exciton liquid. Surprisingly, this exciton liquid is contained in a macroscopic crystal, a solid, nevertheless, which becomes extremely hard due to the exciton liquid and finally exhibits a strange type of superfluid in a two fluid model, where the superfluid phase increases more and more below about 20 K until the whole exciton liquid becomes a superfluid at zero temperature. Never else a superfluid phase has been observed at such high temperatures.

Keywords: excition-polarons, superfluidity, exciton, condensation, heat conductivity, compressibility, sound velocity

1. Introduction

Excitons are electron-hole pairs and as such known in many materials, even in Si. Generally, it is not easy to create so many excitons that they can interact with each other and finally can even condense in an exciton liquid. In standard experiments with laser pulses, one can excite in semiconductors electrons from a valence band into a conduction band and then, due to the electron-hole attraction, the final state of the excited electron drops to somewhat below the bottom of the conduction band. An exciton is thus mobile, but it does not carry an electrical current due to its charge neutrality.

However, if one uses rare-earth compounds, where the uppermost occupied state is a localized 4f state and the lowest conduction band is a 5d band and one makes an optically induced 4f-5d transition, the hole in the 4f state will be localized and the electron just below the bottom of the conduction band will also be localized since it binds by Coulomb attraction to its hole, and the whole exciton will stay at the atom where the photoexcitation occurred and it will not be mobile. Then, this excitation energy will decay in typically 10^{-8} sec with the emission of a photon or phonons at the same atom where it is originated, resulting only in a localized excited atom in the lattice.

However, a p^6–5d exciton is always possible since the initial state is in a band. Thus, Mitani and Koda [1] found Mott-Wannier excitons with thermo-reflectance in Eu chalcogenides with about 4 eV (consult similar TmTe in **Figure 1**). The Tm mono-chalcogenides exhibit a metal-semiconductor transition inasmuch as $Tm^{3+}S^{2-}$ + e is a trivalent metal with one free electron in the 5d conduction band (**Figure 1**). The occupied $4f^{12}$ level is about 6.5 eV below the Fermi energy E_F, and the empty $4f^{13}$ level is little above E_F. Experimental evidence comes mainly from X-ray photoemission spectroscopy (XPS) and Bremsstrahlen isochromat spectroscopy (BIS) [2]. TmTe on the other hand is a divalent semiconductor $Tm^{2+}Te^{2-}$ with an occupied $4f^{13}$ level 0.3 eV below the bottom of an empty 5d band (**Figure 1**). The driving force behind this different character is the crystal field splitting of the 5d band, which depends on the lattice constant due to the different anion radii.

TmSe, on the other hand, with an intermediate anion radius between sulfide and telluride has such a crystal field splitting of the 5d band that the bottom of this band overlaps with the $4f^{13}$ level. This f-d hybridization on the one hand leads to some d-character of the f-state and as a consequence to a narrow f-band and on the other hand to some f-character of the

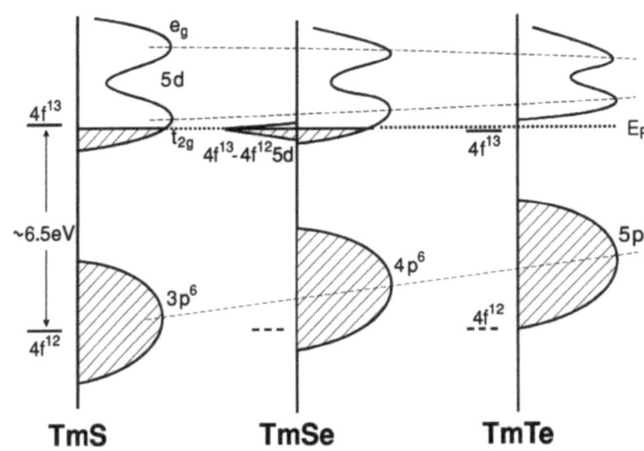

Figure 1. Electronic structure and density of states of the Tm chalcogenides, normalized to the Fermi level E_F.

bottom of the d-band. It has been conventional to describe the new hybridized f-state as $4f^{13}$-$4f^{12}5d$, consisting of a quantum mechanically mixed state [3]. This new phenomenon is called intermediate valence, since the valence of rare-earth ions is defined by the occupation of the f-state, and thus. TmSe has a valence, between 3^+ and 2^+, in fact 2.85^+. This can only be achieved if the 4f-state is a narrow band, which is partially filled with electrons [3]. But also in the 5d band there are some free electrons, which yield in the visible a coupled plasma resonance of these electrons and are responsible for the copper-like color of the crystals.

2. Material tailoring

One can now make mixtures between the semiconducting TmTe and the metallic intermediate valence TmSe and thus tune the energy gap ΔE between 300 meV and zero (metal) [3, 4]. Experimentally semiconducting $TmSe_{1-x}Te_x$ has been created with x = 0.40, 0.55, 0.68, corresponding to ΔE of 40 meV, 110–120 meV, 170 meV. For these compositions the f-state is so close to the 5d band that some hybridization occurs between the tails of the wave functions. We have the unexpected situation of intermediate valence semiconductors. This in turn means that the originally localized $4f^{13}$-state acquires now some bandwidth in the order of tens of meV.

Concerning now the existence of 4f-5d excitons, we have created a situation where the hole state in the 4f is mobile and the electron is in a 5d state below the bottom of the 5d band. Now we have the possibility of 4f-5d excitons. Of course the hole in a narrow 4f-band has a large effective mass, so that the 4f-5d exciton is a heavy ($m_h \approx 50\ m_e$) nearly immobile particle. These excitons without application of external pressure have a low concentration at low temperatures because thermal excitations into the excitonic state are rare.

From Bohr's formula for the hydrogen radius $a_H = 0.53\ \varepsilon$ Å and from **Figure 2** with the reflectivity for $\omega \to 0 = 50\%$, we obtain $n = \frac{1+\sqrt{R}}{1-\sqrt{R}} = 5.8 = \sqrt{\varepsilon}$. So the static dielectric constant is ε = 34, and the radius of the orbit of the exciton is about 18 Å. This would be a Mott–Wannier exciton. Its binding energy from the optical result is theoretically $E_B = 13.6/\varepsilon^2 = 15$ meV, about the same as from the electrical measurement [4], but experimentally from **Figure 2** $E_B \approx 60$ meV. The most complete measurements have been performed on $TmSe_{0.45}Te_{0.55}$ with an energy gap ΔE of 110–120 meV, confirmed by far infrared reflectivity (see inset of **Figure 2**). Similar absorptive peaks as for $TmSe_{0.45}Te_{0.55}$ are absolutely missing in other divalent rare-earth chalcogenides with only localized 4f-states (**Figure 3**).

In the fcc rocksalt structure, the $4f^{13}$-band has a maximum at the Γ point of the Brillouin zone and a minimum at the X point. The 5d band dispersion has its minimum at the X point. An optical transition between the maximum of 4f and the minimum of 5d would be an indirect transition and requires maximal the assistance of a Γ–X phonon for k conservation (black curve).

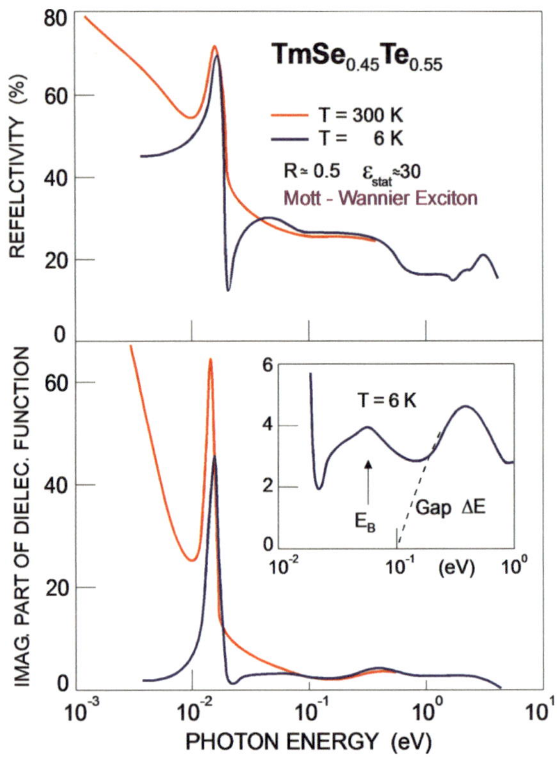

Figure 2. Reflectivity of $TmSe_{0.45}Te_{0.55}$ between 1 meV and 6 eV photon energy. At low temperatures, the transverse optical (TO) phonons are the dominant feature. The inset in **Figure 2** shows the absorptive part of the dielectric function and the energy gap $\Delta E \approx 110$ meV, and the binding energy of an exciton is $E_B \approx 60$ meV [5].

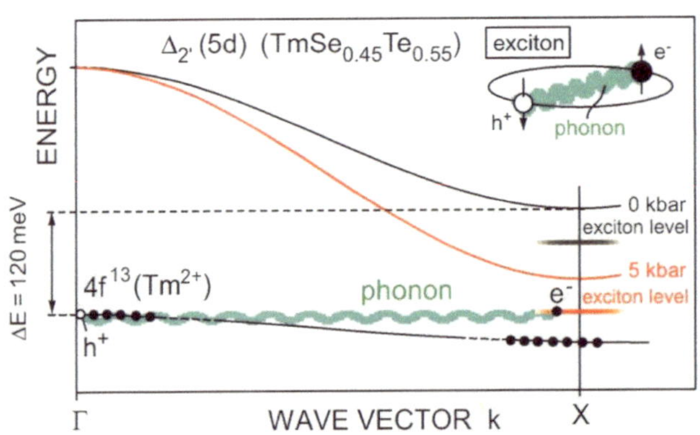

Figure 3. Schematic band structure of $TmSe_{0.45}Te_{0.55}$. Due to 4f-5d hybridization, the $4f^{13}$-state becomes a narrow band and has a dispersion. The exciton level with binding energy E_B is indicated below the bottom of the 5d conduction band (black curve). The red curve represents the band structure at 5 kbar with the exciton level at X at the same height as the 4f-level at Γ. In green is the Γ–X phonon [6].

3. Creation of excitons

Under hydrostatic pressure, the bottom of the 5d band at X with its exciton level will be lowered with respect to its center of gravity ($5dt_{2g}$-$5de_g$) and shown for 5 kbar the exciton level is exactly at the energy of the 4f-state at Γ (red curve). Now the highest energy electrons in the $4f^{13}$-band can spill without energy loss into the excitonic state at X leaving behind a positive hole. This transition needs the emission or absorption of Γ–X phonons which couple to the excitons. So in fact we are dealing with an exciton-polaron. With higher pressure, the bottom of the 5d band at X will approach the energy of the $4f^{13}$-state at Γ and the 4f electrons will enter directly the 5d band and perform a first-order semiconductor–metal transition.

In **Figure 4**, these transitions can be observed directly with resistivity in the isotherms versus pressure for $TmSe_{0.45}Te_{0.55}$. We look at first at room temperature (300 K) and find a classical pressure dependence of a resistivity, namely the resistivity of a semiconductor decreases with increasing pressure, because the energy gap ΔE decreases with pressure and bands widen and finally the metallic state is achieved (above 11 kbar). Starting with about 5 kbar and best observed at 5 K, the resistivity now increases by about three orders of magnitude with pressure. This is exactly the pressure where excitons become stable states and electrons from the f-band, which have been thermally excited into the 5d conduction band, drop into the excitonic state and are no longer available for electric conduction. We have created an excitonic insulator, a term coined by Sir Nevil Mott [7]. With further pressure increase, the resistivity drops again, until now the 4f electrons can enter the 5d band directly, which leads to a first-order semiconductor–metal transition.

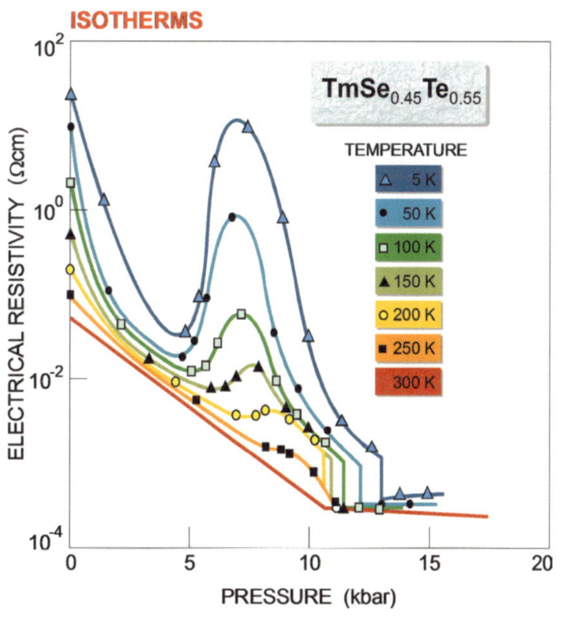

Figure 4. Isotherms of the electrical resistivity in $TmSe_{0.45}Te_{0.55}$ [5].

Here we want to make a remark of another possibility of the semiconductor-metal transition, namely a Mott transition to an electron-hole plasma or an electron-hole liquid. The experimentally derived exciton concentration is 3.9 x $10^{21}cm^{-3}$ (see below). This is in fact too high (because of screening effects) for an electron-hole liquid as has been shown by Monnier et al. [8]. There it is calculated that the electron-hole liquid must be less than $10^{20}cm^{-3}$ excitons. In fact the rare earth nitrides may serve as experimental examples [9].

4. Exciton condensation

Since for the exciton creation no energy is needed, their number is enormous. But not all 4f electrons can form excitons, because as electric dipoles and according to the Pauli principle [10, 11] they repel each other. This goes so far that the formation of this incredible high concentration of excitons forces the whole crystal lattice to expand against the applied pressure. We show this in **Figure 5** where we measure the lattice constant (**Figure 5a**) (with strain gauges) and the expansion coefficient (**Figure 5b**) of the crystal in an isobar at 11.9 kbar. We observe that at about 230 K the lattice expands by 1.6% isostructurally, an enormous amount. The expansion coefficient becomes negative, of course. We even think that the expansion is of first order (dashed–dot line), but the point-by-point measurement cannot reproduce this exactly, because we go from the semimetallic state to the excitonic state.

We can estimate the maximal number of excitons with the help of **Figure 5a**, and we observe that the lattice expansion occurs spontaneously when entering the excitonic phase. There

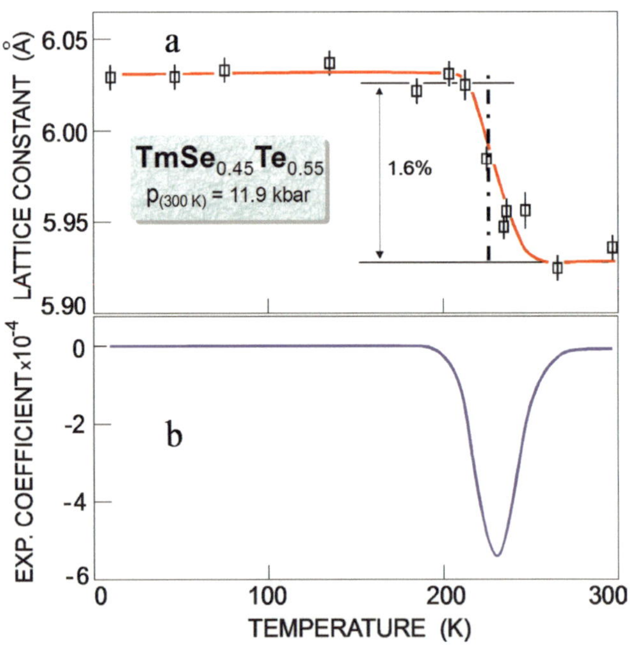

Figure 5. a, b. Isobar lattice constant and expansion coefficient of $TmSe_{0.45}Te_{0.55}$ [12].

http://dx.doi.org/10.5772/intechopen.70095

must be an energy balance between the lattice energy causing the expansion and the electronic energy of the excitons. The energy balance can be described by the first equation in **Figure 6**. We take the lattice constant change from **Figure 5a** to go from 5.93 to 6.03 Å and compute $\Delta l/l$ and $\Delta V/V$. We choose a pressure of 8.5 kbar and an E_B of 70 meV and compute the number of excitons $n_{ex} = 3.9 \times 10^{21}$ cm^{-3} (red field). We also can compute the number of Tm ions in the crystal in the fcc structure, and it is $n_{Tm} = 1.8 \times 10^{22}$ cm^{-3} (yellow field). In other words, the exciton concentration is about 22% of the atomic density, an enormous amount of excitons. With the exciton orbit of 18 Å, it is quite clear that we have an exciton band or an exciton condensation. Since the exciton couples to a phonon, the condensation is a Bose condensation, not a Bose–Einstein condensation. We can also estimate the Bose condensation temperature shown in **Figure 6**, where the general accepted formula yields $T_B = 130$ K, the right order of magnitude. The holes of the exciton are in a narrow 4f-band, and with a pressure change of 5 to 8 kbar (**Figure 4**), one scans the width of the 4f-band [12]. The closing rate of the semiconductor rate has been measured to be $d\Delta E/dp = -11$ meV/kbar [4], so 3 kbar · 11 meV = 33 meV for the width of the narrow 4f-band. From this in turn, we use the

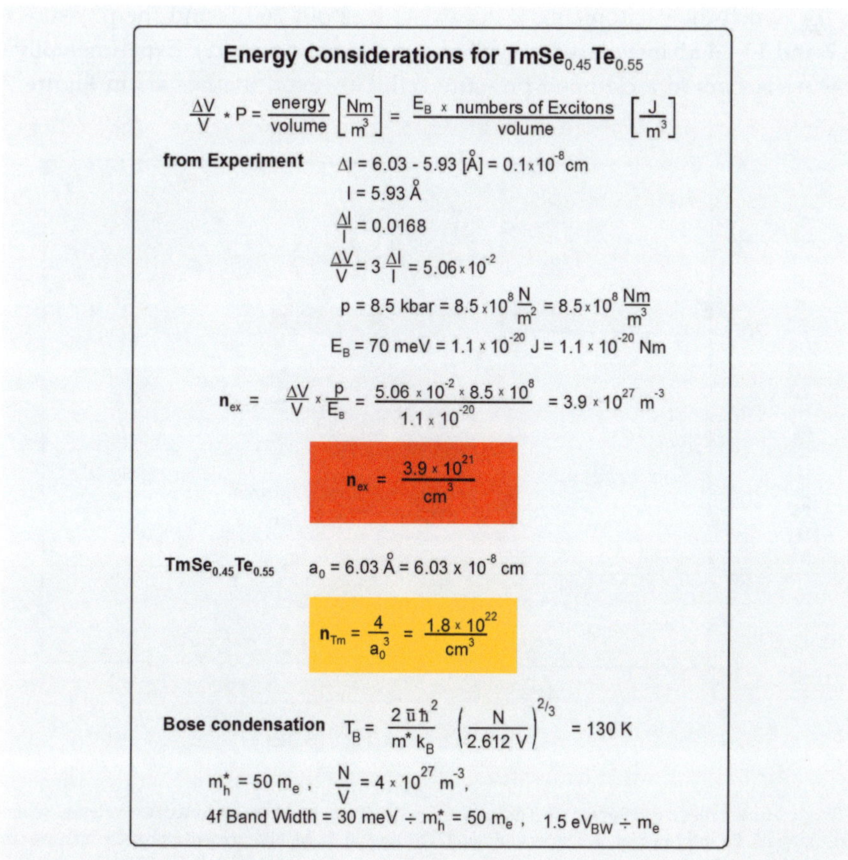

Figure 6. Calculations of the exciton concentration.

general estimate that a band width of 1.5 eV yields an effective mass of m_e and derive that a band width of 30 meV corresponds to an effective hole mass $m_h \approx 50\ m_e$. The excitons are thus heavy bosons.

Here we want to make some remarks about this exciton condensation. Nobody in the world (to the best of our knowledge) has a comparable concentration of excitons which exist as long as we can sustain the pressure and as the liquid Helium lasts, this means for days. We can make all kinds of experiments in this condition, such as electrical conductivity, Hall effect, compressibility, heat conductivity, superfluidity, ultrasound velocity, phonon dispersion and specific heat. Nobody else has these possibilities. But the experiments are very demanding at low temperatures with simultaneous pressure and doing the measurements.

5. Phase diagram of semiconductor, excitonic insulator and semimetal

We plot in **Figure 7** the coexistence ranges of the intermediate valence semiconductor, the excitonic insulator and the intermediate valence semimetal. We see that the highest temperature for which the excitonic insulator exists is about 260 K and the pressure range is between 7 and 13–14 kbar (pressures applied at room temperature). Experimentally one can only measure isobars in a clamped pressure cell. However, the isobars in **Figure 7** are no

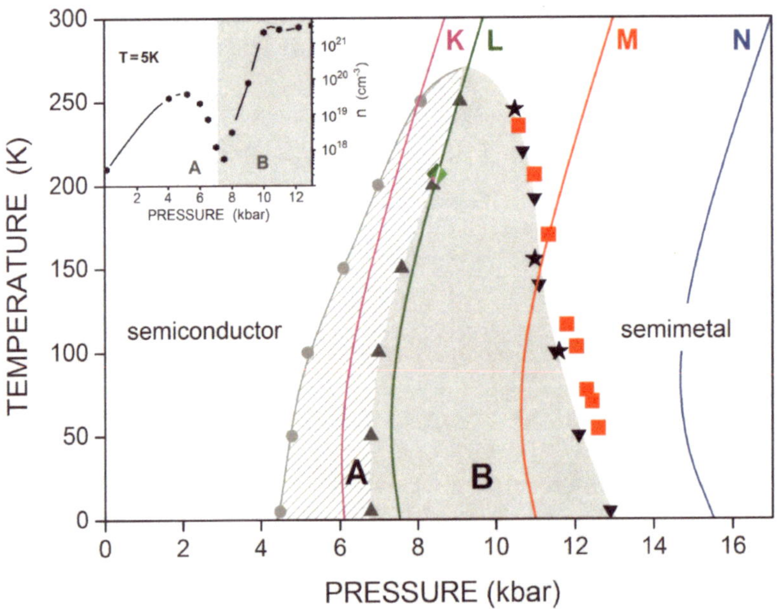

Figure 7. Temperature–pressure diagram of $TmSe_{0.45}Te_{0.55}$ with three regions: intermediate valence semiconductor, excitonic insulator (A, B), intermediate valence semimetal. The lines K, L, M, N represent isobars, which are curved since the pressure applied at 300 K relaxes somewhat at low temperatures. The inset shows the 5d free carrier concentration from a Hall effect in function of pressure and at 5 K [6, 13].

straight lines, because the pressure applied at room temperature relaxes somewhat at low temperatures. In the inset of **Figure 7**, we see the Hall effect, which measures the free electron concentration in the 5d band. In the semimetallic state (curve M) at 13 kbar, the electron concentration is about 3×10^{21} cm^{-3}. For the excitonic insulator at 8 kbar, the free electron concentration is about 10^{18} cm^{-3} because now the free electrons condense into the excitons and do not contribute anymore to the Hall effect. In fact we observe that the carrier concentration reduces by about three orders of magnitude, the same as has been observed in **Figure 4** for the electrical resistivity. The change in resistivity is thus mainly an effect in the carrier concentration though the mobility changes also somewhat [13]. The concentration of the excitons is then 3×10^{21} cm^{-3}–10^{18} cm^{-3} = 3×10^{21} cm^{-3}, about the same as has been obtained in **Figure 4**.

We can consider in an analogy a pot with soup. The pot is the hard surrounding of the crystal and inside is a soup of liquid excitons.

In **Figure 8**, we show in the upper part a proposal from Kohn [14] from 1968 with the three-phase semiconductor, excitonic insulator and semimetal plotted against the energy gap ΔE with increased pressure going to the left. When $\Delta E = E_B$, the excitonic instability starts. In the lower part of the figure, we show the E-k diagram again for the three phases. It is surprising

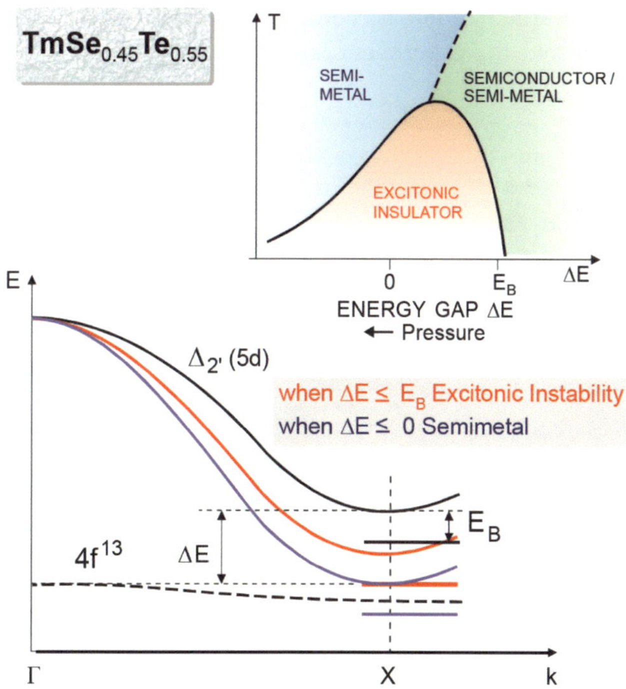

Figure 8. In the upper part, we show a proposal by Kohn [14] of the excitonic insulator, long before any experimental evidence. When $\Delta E = E_B$, the excitonic instability starts. In the lower part, we show again the E-k diagram of $TmSe_{0.45}Te_{0.55}$ in three phases.

and satisfying that the foresight of Walter Kohn has practically reached reality by comparing the inset of **Figure 8** with the real phases of $TmSe_{0.45}Te_{0.55}$ as shown in **Figure 7**.

6. Isotherm and compressibility

We may ask what the direct evidence for the condensed excitonic state is. Typical for any liquid is its incompressibility. We can, for instance, at 1.5 K, apply an increasing pressure to $TmSe_{0.45}Te_{0.55}$, and this is shown in **Figure 9** [12]. At first, we cool at zero pressure from 300 to 1.5 K and volume and lattice constant decrease. Then, we increase pressure and measure the lattice constant with elastic neutrons through the pressure cell. Of course, lattice constant and volume decrease further, corresponding to a Birch–Mournaghan equation (red curve). This is a very time-consuming experiment, because for each pressure change the pressure cell had to be heated to room temperature to change to a higher pressure and then cooled down again and adjust the sample in the neutron beam and wait for beam time. Therefore, this experiment has only four points, but at the relevant pressures. As can be seen in **Figure 9** when entering the excitonic state, the lattice constant remains unchanged with increasing pressure, which means a compressibility of zero, as shown in **Figure 10**.

Taking experimental uncertainties into account, we have at least a compressibility just as for diamond. Thus, we can take this experiment as evidence of an excitonic liquid.

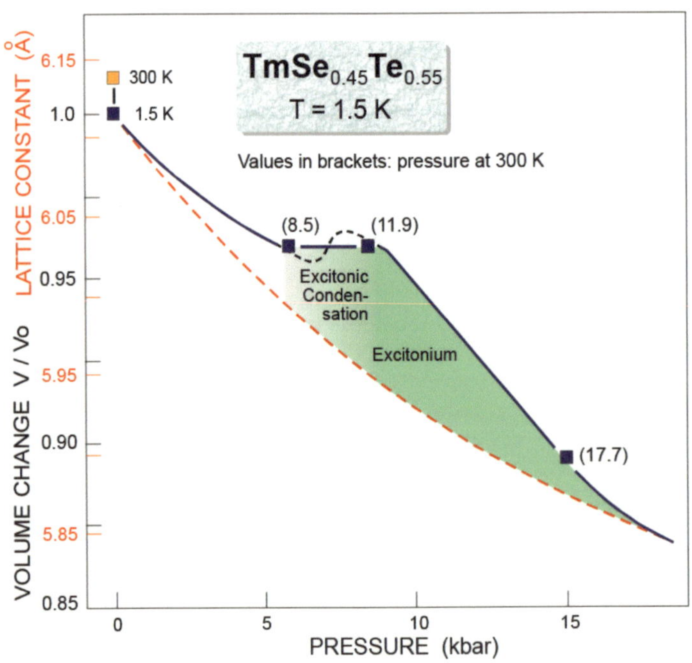

Figure 9. Isotherm of $TmSe_{0.45}Te_{0.55}$ at 1.5 K at relevant pressures. In brackets values at 300 K [12].

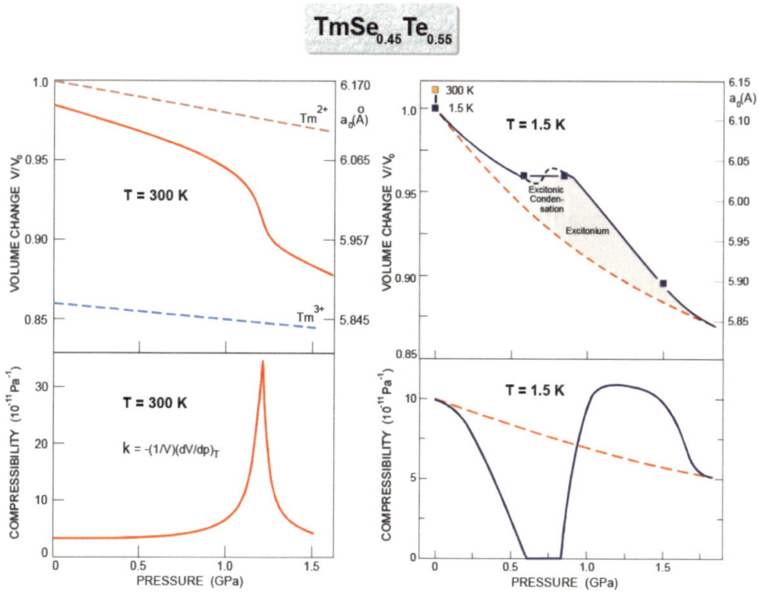

Figure 10. At room temperature (left-hand figures), the volume change with pressure has a dramatic change near 11 kbar at the transition semiconductor-semimetal, because the material is intermediate valence between Tm^{2+} and Tm^{3+} (also shown as reference in **Figure 10**), and it becomes soft with pressure. On the right-hand side, we show the material at 1.5 K and the compressibility goes to zero; the material becomes extremely hard, because we have now the exciton condensation.

7. Heat conductivity and superfluidity in the excitonic liquid

We now want to discuss the possibility of superfluidity in the excitonic liquid. Here we resort at first to theory [11, 15]. There is a similarity between pairs of particles: two electrons can condense and produce superconductivity, and an electron–hole pair (exciton) can upon condensation result in superfluidity. A positron pair should also result in superconductivity, but no such experiment is known.

In any case, our exciton condensation may result in superfluidity. What would be the experiment to prove this? In our opinion, this is heat conductivity [16], because it would diverge when the material becomes superfluid, just as in ^4He [17].

The experimental arrangement to measure heat conductivity and thermal diffusivity in a pressure cell is described in detail in Ref. [16], but the essence is isobars between 4 and 300 K at various pressures. We show the results of measurements of the heat conductivity λ with isobars at four different pressures, one in the semiconducting range (compare **Figure 7**) with 7 kbar, one in the semimetallic range at 15 kbar, both outside the excitonic region, and at two pressures 13 and 14 kbar within the excitonic range. Temperature has been measured automatically for each degree. In **Figure 11**, we collect a few relevant formulae for the heat conductivity λ and the thermal diffusivity a. We see that the heat conductivity depends on the specific heat c_v and l_{ph} in direction x, the mean free path

Heat Conductivity λ and Thermal Conductivity (Diffusivity) a

from kinetic gas theory $\lambda = 1/3\, C_v\, v\, l$

with temperature gradient dT/dx $\lambda = \rho c_v v_x l_x$

c_v = spec. heat/kg
l_x = mean free path
v_x = particle velocity
ρ = density

$c_v = \lambda/\rho v_x l_x \quad \lambda/\rho v_x^2 \tau = \lambda/\rho a$ with $a = v_x l_x = v_x^2 \tau$

for $T > \Theta_D$ (Dulong-Petit) $c_v = 3R$ $\lambda = \lambda_{ph} + \lambda_{ex} + \lambda_{el}$

for metal: $\lambda_{ph} + \lambda_{ex} << \lambda_{el}$

for insulator: $\lambda_{el} << \lambda_{ph} + \lambda_{ex}$

Figure 11. Formalities for the heat conductivity and the thermal diffusivity.

for phonon scattering. In short, l_{ph} will increase with decreasing temperature because the density of phonons decreases and we have Umklapp processes involving three phonons. But the specific heat c_v definitely will go toward zero for zero temperatures; thus, the heat conductivity outside the excitonic region will display a maximum near 50 K, as well for the semiconducting range (7 kbar) as for the metallic range (15 kbar); this is displayed in **Figure 12**, and this behavior is quite normal. The difference of the heat conductivity near 300 K for both cases is due to the electronic part of the heat conductivity in the metallic state, and it corresponds roughly to the Wiedemann–Franz relation. This gives confidence to the measurements. We continue with the heat conductivity in the excitonic region at 13 and 14 kbar. We observe an unexpected downward jump in a first-order transition when entering the excitonic phase. Consulting **Figure 7**, it is obvious that at different pressures one enters the excitonic phase at different temperatures. At these temperatures and pressures, one enters the insulating excitonic phase mainly from the semimetallic phase, thus with a metal–insulator transition. The downward jumps in the heat conductivity λ reflect the loss of the electronic part of the heat conductivity. The fascinating aspect of the heat conductivity in the excitonic region is the sharp increase of λ below about 20 K, quite in contrast with the λ outside the excitonic region. Since λ follows mainly the specific heat c_v and the phonon mean free path l_{ph}, (**Figure 11**) and c_v nevertheless must go to zero for $T\rightarrow 0$, it is the phonon mean free path which goes faster to infinite than c_v toward zero. Finally, it means that the phonon mean free path becomes

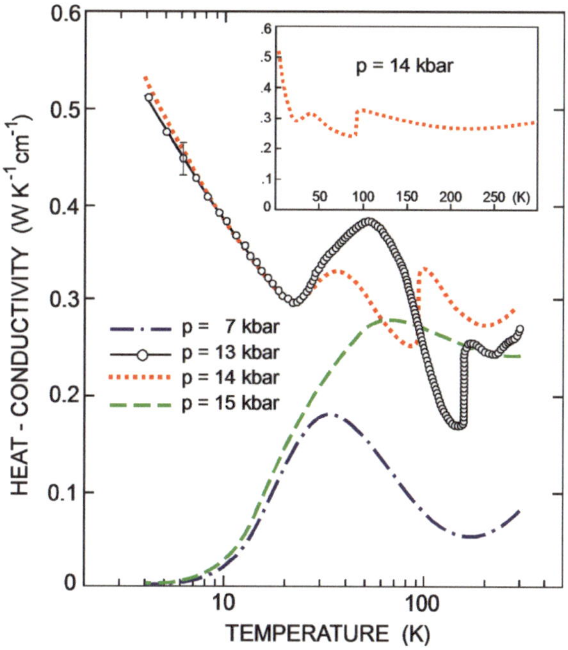

Figure 12. Heat conductivity λ of $TmSe_{0.45}Te_{0.55}$ for various pressures in function of temperature. Dotted and full line in the excitonic region, dashed in the semimetallic region and dash–dotted line in the semiconducting phase. The inset shows the heat conductivity at 14 kbar in a linear scale [16].

infinite. When one makes a heat pulse at one end of the crystal, the excited phonon transports its energy without scattering on other phonons to the other side of the crystal, meaning an infinite heat conductivity. This is, however, only possible if the concentration of phonons as running waves is substantially reduced, because most of them couple to the heavy excitons as exciton-polarons, as we have seen before and thus more or less correspond to local modes.

Unfortunately, the measurements were limited to 4.2 K, because at the time of the measurements one did not realize the implications. In any case 20 K, the onset of the sharp increase of λ with decreasing temperature can be considered as the onset of superfluidity, which, however, is different from the one of ^4He, inasmuch as there the onset of superfluidity is a first-order transition [17]. For our exciton case, we propose a superthermal current in the two-fluid model, where the superfluid part increases gradually toward zero temperature [17]. A λ-anomaly in the specific heat as in the first-order Bose–Einstein transition in ^4He is here not to be expected and also not found [17].

The proposed evidence of superfluidity within the condensed excitonic state necessitates an additional excitation spectrum of other quasiparticles, namely rotons or vortices [17]. λ_{tot} is the sum of individual contributions (**Figure 11**), and below about 20 K $\lambda_{tot} = \lambda_{ph} + \lambda_{ex}$. λ_{ph} is the heat conductivity due to uncoupled phonons, which is proportional to T^3 and can be neglected compared to λ_{ex} at low temperatures. Thus, we obtain for λ_{ex} an Arrhenius law for the increase of the heat conductivity toward zero temperature $\lambda_{ex} \propto \exp \Delta/k_BT$. This is shown

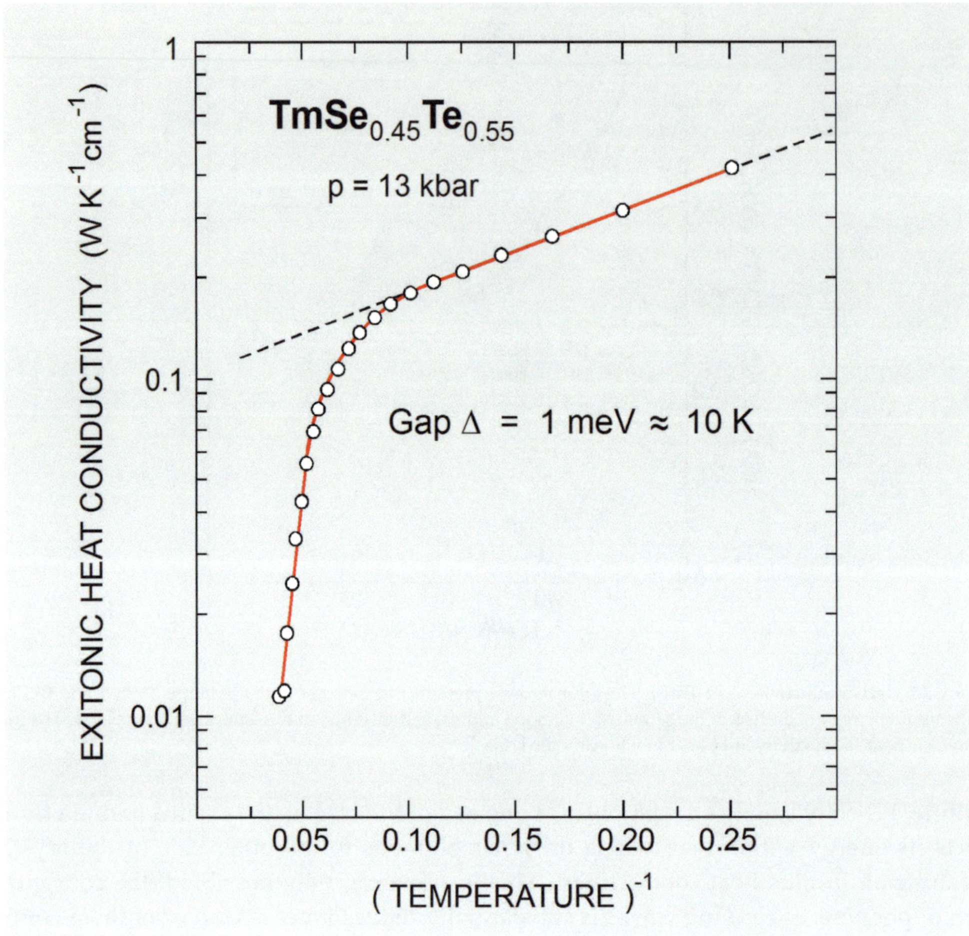

Figure 13. Excitonic part of the heat conductivity λ_{ex} at 13 kbar, shown in an Arrhenius plot [16].

in **Figure 13** at 13 kbar. The activation energy or the gap Δ is 1 meV or about 10 K. The application of heat in the heat conductivity experiment can excite quasiparticles, e.g., rotons with gap energy of about 5 K, which is the right order of magnitude. In superfluid ^{4}He, the roton gap is 8.65 K [18].

8. Thermal diffusivity

In **Figure 14**, we display the thermal diffusivity a for the same four pressures as in **Figure 12**. The thermal diffusivity $a = v_x l_x$ and thus follows mainly the phonon mean free path l_x, with v_x being about constant outside the excitonic region, consulting **Figure 11**. In fact outside the excitonic region with 7 and 15 kbar, it does exactly this, as can be seen in the theoretical curve for l_x in **Figure 15**. For the lowest temperatures, the phonon mean

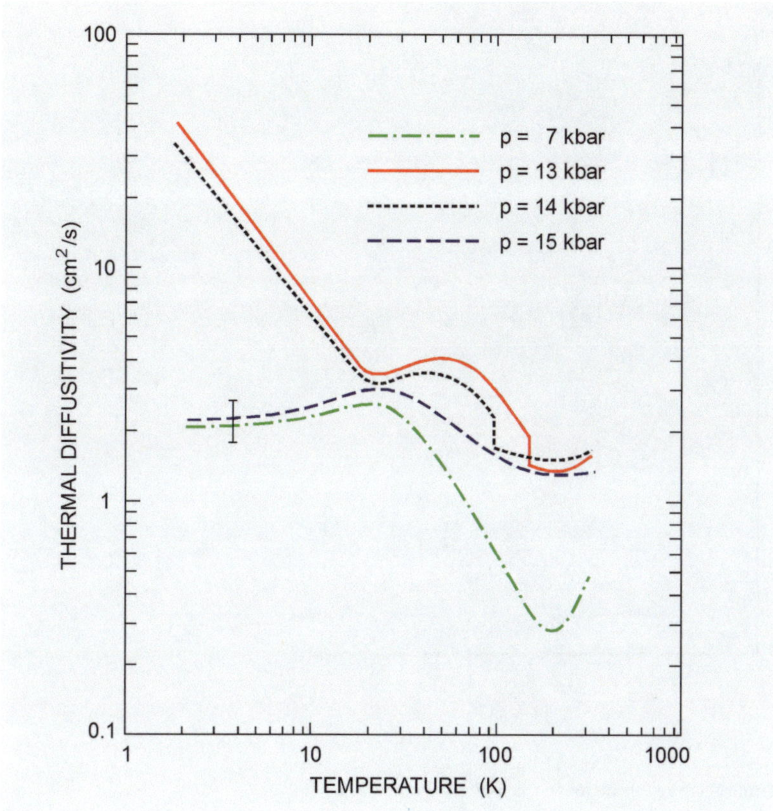

Figure 14. The thermal diffusivity in the semiconducting (7 kbar), semimetallic (15 kbar) and the excitonic phase (13 and 14 kbar) for $TmSe_{0.45}Te_{0.55}$ [16].

free path becomes the dimension of the crystal and is thus constant. But in the excitonic region, again below about 20 K, the thermal diffusivity increases dramatically. Why then in the excitonic region the dimensions of the crystal do not seem to be important now? Just as in superfluid Helium heat can be transferred not only via phonon-phonon scattering in a diffuse manner, but ballistically via a highly directional quantum mechanical wave, the second sound. Also above 20 K, there are anomalies, but they can be explained with the velocity of sound v_x, which increases now strongly in the excitonic region (see below).

In principle, the two measurements of heat conductivity and thermal diffusivity permit the calculation of the specific heat $c_v = \lambda/\rho a$, with ρ the density (see **Figure 11**) [16] and we did this in **Figure 7** of Ref. [16]. But we never felt very happy with this curve because we divided two point-by-point measurements. But the specific heat in the excitonic range is definitely below the one of the specific heat outside this range. But it is also very complex since the density ρ diminishes when entering the excitonic phase, because the crystal expands (see **Figure 5**). It took us several years before we could make a direct measurement of the specific heat under pressure and below 300 K [6, 19].

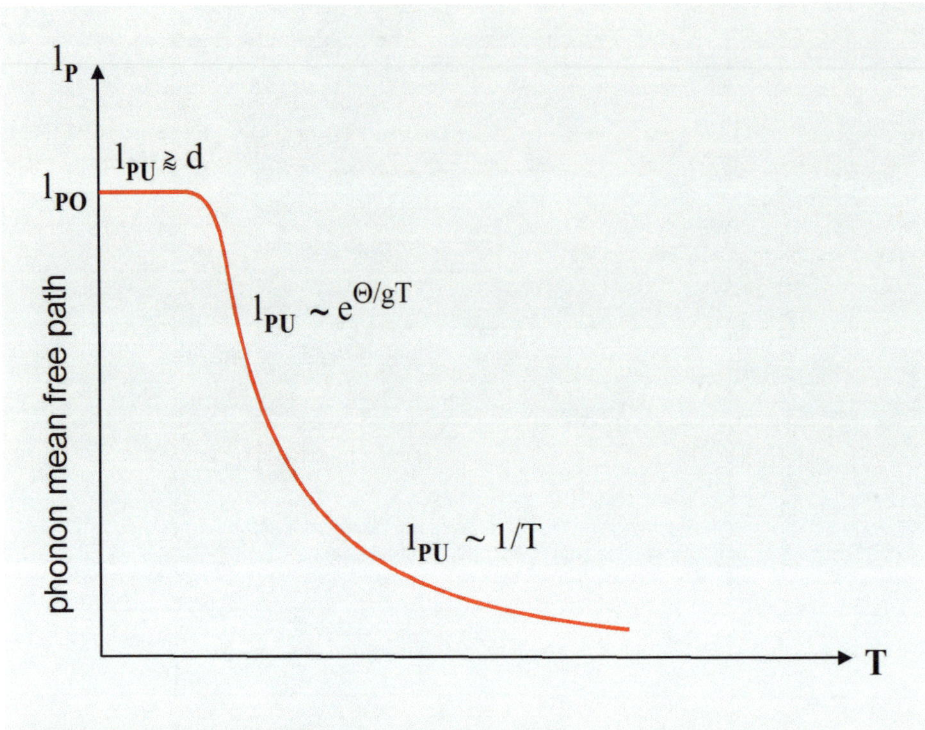

Figure 15. Theoretical curve for the mean free phonon path.

9. The specific heat

The specific heat c_v has been measured for $TmSe_{0.45}Te_{0.55}$ [6, 19] along isobars with 0 kbar, and corresponding to the curves K, N, M in **Figure 7**, which is shown in **Figure 16** [6, 19].

The molar specific heat c_m in J/mole K/f.u. has been matched at 300 K to the Dulong-Petit value of 52 J/mole K/f.u. The specific heat at ambient pressure represents a normal Debye curve (black curve). This curve has been measured by our colleagues at the university of Geneva and ETH Zürich for T > 1 K and T > 0.3 K [20]. A Schottky anomaly due to crystal field splitting of the Tm ions and an exchange splitting due to magnetic order at 0.23 K has been subtracted from the measured curve, and the pure phonon contribution could be plotted as c_v/T versus T^2; thus, a Debye temperature Θ of 117 K could be obtained [16]. Curve N is in the semimetallic high-pressure phase outside the excitonic region, and we find again a normal Debye curve (blue curve), but with a lower Debye temperature than at ambient pressure. This is at first sight surprising since at high pressure a solid becomes harder with a higher Debye temperature, but it has also been observed in reference [16]. A simple explanation can be that with high pressure we change somewhat the degree of valence mixing in the intermediate valence semimetal. Curve M (red curve) starts with about 13 kbar at 300 K in the metallic region, but enters the excitonic region at about 150 K. This occurs with a first-order transition as we see in **Figure 5a**.

http://dx.doi.org/10.5772/intechopen.70095

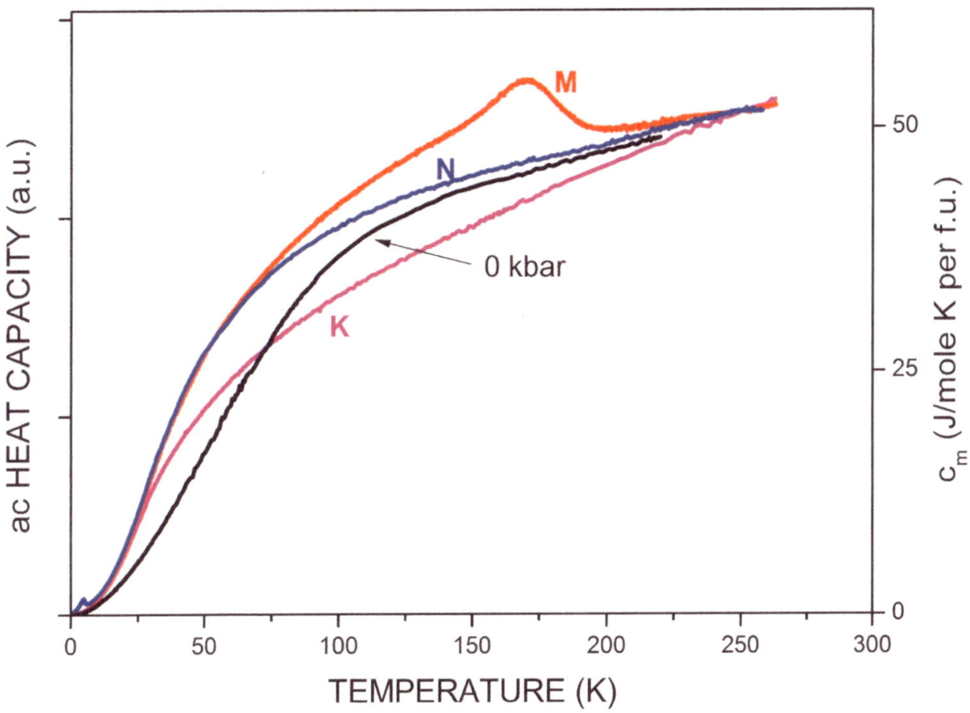

Figure 16. The measured specific heat of $TmSe_{0.45}Te_{0.55}$ at various pressures. The colors and letters are the same as in **Figure 7** [6].

The specific heat should reveal a delta-function at the phase transition, but experimentally the spike reduces to a Gaussian shape [21]. All curves entering the excitonic region in **Figure 7** from the semimetallic region in a first-order transition (red squares and downward triangles in **Figure 7**) exhibit the Gaussian anomaly. We now discuss curve K (mauve), which enters the excitonic region in a second-order transition without a spike in the specific heat. We observe that the specific heat is no longer a Debye curve, but below about 250 K (arrow in **Figure 17**) one finds a quasi-linear drop of the specific heat until below about 30 K the specific heat joins the other curves. These measurements reveal a fundamental difference of the thermodynamic phases A and B in **Figure 7**.

In fact, such a specific heat like curve K with a non-Debye like curvature has never been seen before. Since the specific heat over a higher and larger temperature interval is entirely given by the phonons (in the absence of magnetic order and special effects like Schottky anomalies), we must conclude a strong renormalization of the phonon spectrum.

As we have stated already several times above the excitons in this indirect semiconductor couple strongly to phonons in a triple particle entity of hole-electron and phonon as an exciton polaron. But when the phonons couple to the heavy excitons with effective masses of the holes around $m_h = 50\ m_e$ they become more or less localized like a local mode and do no longer contribute significantly to the specific heat. So an essential part of the Debye spectrum of the specific heat is missing.

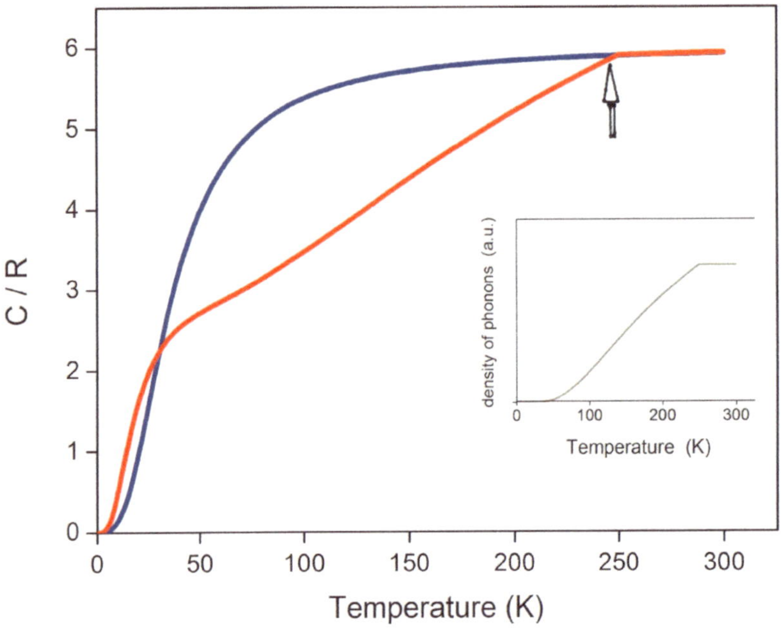

Figure 17. The specific heat of two typical schematic curves. One representing curve N in **Figure 7** and typical for a Debye curve. The second representing curve K in **Figure 7**. The inset shows an assumed linear temperature dependence of the optical phonon density of states [6].

So in **Figure 17** we have made a model calculation of the specific heat with the assumption that the optical phonons are bound to the excitons below about 250 K (arrow in **Figure 17**). The acoustic branches of the phonons are modelled with a Debye and the optical phonons with an Einstein ansatz, respectively [6]. The seeming disappearance of phonons, i.e., the binding of the optical phonons on the excitons is represented with a linear decrease of the density of states below 250 K (inset to **Figure 17**). The model calculation in **Figure 17** represents well the intriguing behavior of curve K in **Figure 16**. A strong coupling regime for the phonons to excitons prevails, and thus, while cooling, more and more wave-like phonons become locked onto the excitons, giving no more contribution to the specific heat. This renormalization of the phonon spectrum and the resulting effect on the specific heat has never been seen before, and it is due to the extreme large concentration of exciton–polarons. Regarding now the Debye temperature Θ of curve K in comparison with curve N, we observe a further reduction of the Debye temperature, i.e., a minimum in the Debye temperature versus pressure in the excitonic region (red curve in **Figure 17**).

In a quantitative formula, we can express $c_{exp}(T) = c_{ac}(T) + \rho_{opt}(\omega, T)\frac{(\Theta_E/T)^2 e^{(\Theta_E/T)}}{(e^{(\Theta_E/T)} - 1)^2}$

(Debye) + (Einstein) with ρ_{opt} a temperature-dependent density of optical phonons (see inset **Figure 17**). In fact, the model calculation in **Figure 17** represents quite well the measured specific heat of curve K in **Figure 16**. Thus, the acoustic phonons alone exhibit a Dulong-Petit value c/R of 3 cal/degree and the optical phonons have a temperature-dependent density, their decrease with temperature representing the increase of excitons–polarons with decreasing temperature. The free optical phonons get lost for, e.g., the thermal conductivity. However, below about 20 K the excitons–polarons take over in the heat conductivity or the thermal diffusivity and with a diverging increase in these entities finally lead to superfluidity.

Curve M in **Figure 16** is on the decreasing branch of exciton concentration (see **Figure 7**) where more and more free electrons in the 5d band are screening the Coulomb interaction between electron and hole. In dissolving the excitons in region B of **Figure 7**, the electrons from the excitons enhance the 5d electrons, further which leads to a cumulative process and a collective breakdown of the rest excitons in a first-order transition. Curve M enters the excitonic region at a temperature of about 150 K where the exciton concentrations are already about two orders of magnitude lower than at the maximum.

10. Sound velocity

An ultrasound transducer has been glued to one end of the crystal, and with a multiple echo from the other end of the crystal over the known length of the crystal, the sound velocity could be obtained. This is shown in **Figure 18** for various pressures. At zero pressure and at 7 and 18 kbar, the sound velocity is about 4000 m/s and there is not much change with pressure. But best seen at 12 kbar, when entering the excitonic phase at 180 K, the sound velocity is enhanced by nearly a factor 2 (see **Figure 7**). With 10 kbar, we are entering the excitonic phase at 240 K, again with a jump of nearly a factor 2 but near 90 K the pressure loss in the cell was just the size for a reentrant transition to the non-excitonic phase. This was a unique phenomenon, but supporting the experimental measurements.

As mentioned above in the chapter about the thermal diffusivity a, the upwards jumps in the excitonic region are indeed caused by the jumps in the sound velocity. But not only this, the increase in sound velocity contributes directly to the thermal conductivity above 20 K and is responsible for the bumps in the thermal diffusivity.

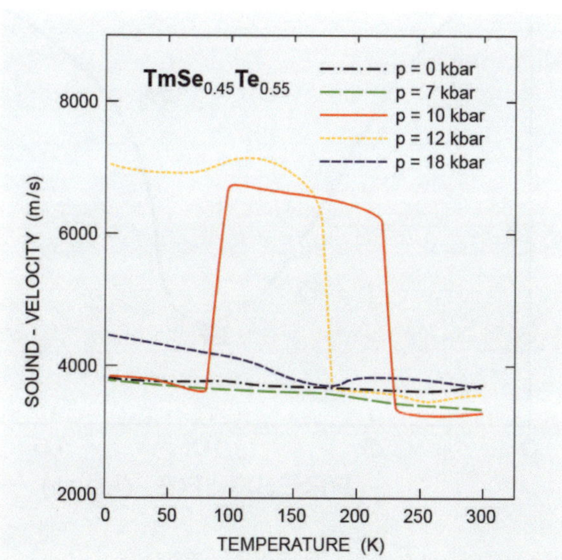

Figure 18. Longitudinal sound velocity outside and inside the excitonic region.

The sound velocity is related to the bulk modulus B and its inverse the compressibility. For a cubic material B depends on the elastic moduli c_{ij} as $B = 1/3\ (c_{11} + 2c_{12})$ the elastic moduli instead, depend on the sound velocity as $c_{11} = \rho v_{L\ [100]}^2$ and $c_{12} = \rho(v_{L\ [100]}^2 - 2v_{T2\ [110]}^2)$. Assuming that in general v_L is about 3 times v_{T2} we get the simplified relation $B \approx \rho v_L^2$. Thus in the excitonic phase we find a 2 times larger v_L and thus a 4 times larger B or a 4 times smaller compressibility. The material gets indeed appreciable harder in the excitonic state.

In **Figure 9**, we have shown that between 5 and 8 kbar the lattice constant remained practically constant during exciton condensation, meaning that the compressibility is close to zero. Putting a maximal error bar through the points of measurement a bulk modulus B = 20 GPa outside the excitonic region and a bulk modulus B = 70 GPa in the excitonic region could be obtained. From the sound velocity measurement in **Figure 18**, we calculate a bulk modulus B = 24 GPa outside the excitonic region and one of 100 GPa in the excitonic region. So both types of measurements agree reasonable well and confirm the fact that during exciton condensation the material becomes extremely hard.

We offer two explanations for this phenomenon: the electron from the exciton enters a 5d-like orbit, which is much larger than the original 4f orbit it came from, and this in spite of the increasing pressure. Or the excitons, being electric dipoles, repel each other at short distances and large concentrations, creating a counter-pressure to the applied pressure.

The dominant feature in **Figure 18** is the sharp increase by a factor two of the sound velocity when entering the excitonic phase. But also at 300 K in an isotherm taken from **Figure 18** with the relevant pressures the sound velocity is changing. Now at 300 K this is shown in **Figure 19**,

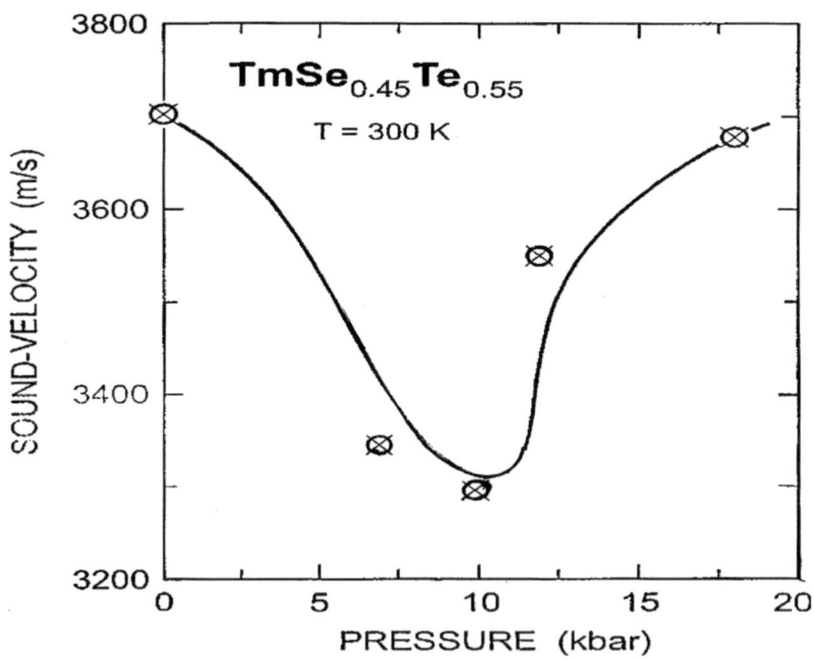

Figure 19. Sound velocity measurements at 300 K as a function of pressure [16].

but it has nothing to do with excitons and the change is much smaller than the one due to excitons. Here we find a minimum of the sound velocity with increasing pressure, inverse to what we have discussed in the excitonic region. The relation of bulk modulus B with sound velocity is $B \approx \rho v_L^2$, and a minimum of sound velocity implies a minimum in the bulk modulus, which is inverse to the compressibility. Thus, the minimum in the sound velocity means a maximum in the compressibility. This can be compared with the compressibility for 300 K in **Figure 10**, and we obtain a similar curve. At 300 K, the softening of the bulk modulus or a maximum in the compressibility is here due to a change of the degree of valence mixing with pressure. So these completely different experiments (also by different authors [4]) support each other and give again confidence into the experiments.

11. Dispersion of exciton-polarons

We mentioned above regarding **Figures 16** and **17** that the Debye temperature in the excitonic region is less than the Debye temperature Θ = 117 K at 300 K but in **Figure 18** we observe that the sound velocity is enhanced in the excitonic region. This seems to be a contradiction since in the Debye model the sound velocity is the slope of a linear phonon dispersion curve where the maximum frequency ω_{max} determines the Debye temperature Θ. A lower Debye temperature has thus a lower sound velocity and a lower bulk modulus. In order to explain a lower ω_{max} together with a higher sound velocity, we have to leave the simple Debye model for bare longitudinal acoustic (LA) phonons $\omega \propto \sin(ka/2)$ and use a new dispersion curve of an exciton–polaron quasi-particle. This is no longer a simple sinus function. The dispersion of such an exciton–polaron is treated in textbooks, e.g., [22]. The result is that the phonon spectrum will be greatly renormalized in the excitonic region [23]. We show in **Figure 20** a LA phonon in Γ-X direction with $\omega_{LA\,(\Gamma\text{-}X)} \approx$ 14 meV [12], in **Figure 20b** an exciton with 4f character $E_B - \hbar^2k^2/2M$, where M is the sum of electron and hole mass of the exciton $M = m_h + m_e \approx m_h$ with $m_h \approx 50\, m_e$.

We can see in **Figure 20** that now the dispersion of the exciton-polaron has indeed simultaneously a steeper slope (larger sound velocity) than the LA phonon and a lower ω_{max} than the simple phonon (smaller Debye temperature).

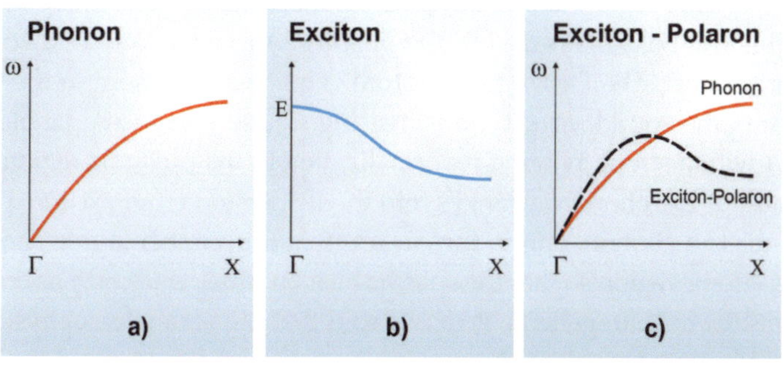

Figure 20. a. Dispersion of a LA phonon. b. Dispersion of a 4f exciton. c. Dispersion of an exciton–polaron.

12. Theoretical Models

Since the binding energy E_B of the exciton-polaron is with 60-70 meV relatively large also a Frenkel type of exciton-polaron is conceivable. Thus in a theoretical paper [24] it is proposed that the exciton condensation occurs in an extended Falikov-Kimball model [25] where, instead of the original model with localized 4f states a narrow hybridized 4f band is used, which is more realistic in this case. Extensively discussed has also been an effective mass model [26] with large differences between electron and hole mass, just as we proposed above. In a further paper [27] it has been shown, that weakly overlapping Frenkel type excitons can condense. Especially the coupling of excitons with phonons has been discussed in Ref. [28] and the formation of exciton-polarons. Finally in [29] exciton densities and superconductivity (sic) are discussed where for low exciton densities a Bose-Einstein condensate is proposed and for high density a Bardeen-Cooper-Schrieffer condensate should prevail, especially in coupled bilayers. This certainly is not the case in our experiments.

13. Conclusion

In this review paper, we treat a special rare-earth material, $TmSe_{0.45}Te_{0.55}$ which has been tailored so that with moderate pressures (up to 20 kbar) and low temperatures (down to 4 K) an enormous amount of excitons (10^{21} cm^{-3}), about 22% of the atomic density, can be statically obtained. This high concentration of excitons with Bohr orbits of about 18 Å leads to a condensation of excitons, which forms a liquid inside a crystalline surrounding. The existence range of condensed excitons is below 250 K and between 5 and 14 kbar. The condensation is accompanied with a phenomenon of incompressibility and as such with a compressibility near zero. In this condition, the heat conductivity and the thermal diffusivity have been measured in order to investigate a possible superfluidity which has been proposed by Keldysh and Kopaev [11] and Kozlov and Maksimov [15]. Outside the excitonic region, both entities behave quite normal, whereas in the excitonic region the heat conductivity diverges to ever-increasing values. This can be explained below 20 K within a two-fluid model, where the superfluid part always increases until at temperature zero the complete condensed excitons become superfluid. Also the thermal diffusivity expands in the excitonic region above the phonon mean free path corresponding to the size of the crystal. This can be explained with the quantum-mechanical second sound, which is a ballistic transport of heat. The Debye temperature exhibits a minimum in the excitonic region where nevertheless the sound velocity is increasing. These two incompatible measurements can be explained with a strong phonon renormalization in the excitonic region, and the Debye phonon dispersion of LA phonons changes into the dispersion of an exciton–polaron, because every exciton binds to a phonon. This in turn means that the number of free phonons is strongly reduced in the excitonic region so that the specific heat becomes extremely anomalous, far away from a Debye specific heat. In general, it can be said that the anomalous physical properties of condensed excitons are unprecedented.

http://dx.doi.org/10.5772/intechopen.70095

$TmSe_{0.45}Te_{0.55}$ is not the only material where these phenomena can be observed. YbO and YbS are similar materials though one will need much larger pressures to close their gaps of about 1 eV [30]. But also $Sm_{0.75}La_{0.25}S$ [31] is a possible candidate for which much lower pressures are needed. So with a good feeling for materials new and exciting effects can be found.

Acknowledgement

The author wishes to thank Prof. Benno Bucher of HSR, Switzerland, for rechecking the experimental results and fruitful discussions.

Author details

Peter Wachter

Address all correspondence to: wachter@solid.phys.ethz.ch

Laboratorium für Festkörperphysik, ETH Zürich, Zürich, Switzerland

References

[1] Mitani T, Koda T. Proceedings of the 12th Int. Conf. Phys. Semiconductors, Stuttgart, Teubner publ. 1974; p. 889

[2] Wachter P, Kamba S, Grioni M. Physica B. 1998;**252**:178

[3] Wachter P: In: Gschneidner jr. KA, Eyring L, Lander GH, Chopin GR. Editors. Handbook on the Physics and Chemistry of Rare Earths, 19 Lanthanides/Actinides: Physics II. 1991. p. 177

[4] Boppart H, Wachter P. Proceedings of Material Research Society. 1984;**22**:341

[5] Neuenschwander J, Wachter P. Physical Review B. 1990;**41**:12693

[6] Wachter P, Bucher B. Physica B. 2013;**408**:51

[7] Mott N. Philosophical Magazine. 1961;**6**:287

[8] Monnier R, Rhyner J, Rice TM, Koelling DD. Physical Review B. 1985;**31**:5554

[9] Wachter P. Advances in Material Physics and Chemistry. 2015;**5**:96

[10] Halperin BI, Rice TM. Reviews in Modern Physics. 1968;**40**:755

[11] Keldysh LV, Kopaev AN. Soviet Physics Solid State. 1965;**6**:2219

[12] Wachter P. Solid State Communication. 2001;**118**:645

[13] Bucher B, Steiner P, Wachter P. Physical Review Letters. 1991;**67**:2717

[14] Kohn W. In: Many Body Physics, edited by de Witt C, Balian R. New York: Gordon & Breach; 1968

[15] Kozlov AN, Maksimov LA. JETP. 1965;**21**:790

[16] Wachter P, Bucher B, Malar J. Physical Review B. 2004;**69**:094502

[17] Tilley DR, Tilley J. In: Superfluidity and Superconductivity. edited by Brewer DF. Bristol and New York: Adam Hilger; 1990. P. 35

[18] Henshaw DG, Woods ADW. Physical Review. 1961;**121**:1266

[19] Bucher B, Park T, Thompson JD, Wachter P. arXiv:0802.3354v2

[20] The authors are most grateful to Prof. A. Junod (University of Geneva) and CH. Bergeman ETH Zürich for performing specific heat measurements at ambient pressure

[21] Schilling A, Fischer RA, Phillips NE, Welp U, Kwok WK, Crabtree GW. Physical Review Letters. 1997;**78**:4833

[22] Hodgson JN. In: Optical Absorption and Dispersion in Solids. London: Butler and Tanner; 1970

[23] Wachter P, Bucher B. Physica Status Solidi ©. 2006;**3**:18

[24] Ihle D, Pfafferot M, Burovski E, Bronold FX, Fehske H. Physical Review B. 2008;**78**:193103

[25] Falikov LM, Kimball JC. Physical Review Letters. 1969;**22**:997

[26] Bronold FX, Fehske F. Physical Review B. 2006;**74**:165107

[27] Zenker B, Ihle D, Bronold FX, Fehske H. Physical Review B. 2012;**85**:121102

[28] Zenker B, Fehske H, Beck H. Physical Review B. 2014;**90**:195118

[29] Soller H. Journal of Applied Mathematical and Physics. 2015:1218

[30] Wachter P. Journal of Alloys and Compounds. 1995;**225**:133

[31] Wachter P, Jung A, Pfuner F. Physical Letters A. 2006;**359**:528

Chapter 3

Theoretical Insights into the Topology of Molecular Excitons from Single-Reference Excited States Calculation Methods

Thibaud Etienne

Additional information is available at the end of the chapter

http://dx.doi.org/10.5772/intechopen.70688

Abstract

This chapter gives an introduction to qualitative and quantitative topological analyses of molecular electronic transitions. Among the possibilities for qualitatively describing how the electronic structure of a molecule is reorganized upon light absorption, we chose to detail two of them, namely, the detachment/attachment density matrix analysis and the natural transition orbitals strategy. While these tools are often introduced separately, we decided to formally detail the connection existing between the two paradigms in the case of excited states calculation methods expressing any excited state as a linear combination of singly excited Slater determinants, written based on a single-reference ground state wave function. In this context, we show how the molecular exciton wave function plays a central role in the topological analysis of the electronic transition process.

Keywords: excited states, excitons, detachment/attachment, transition matrix and orbitals, charge transfer

1. Introduction

Providing a quantitative insight into light-induced electronic structure reorganization of complex chromophores remains a challenging task that has attracted a substantial attention from theoretical communities in the past few years [1–15]. Indeed, a potential knowledge related to the ability of a chromophore to undergo a charge transfer caused by photon absorption or emission [16, 17] is of seminal importance for designing novel dyes with highly competitive optoelectronic properties [18–21]. Most often, such quantitative probing of the charge transfer locality is accompanied by a qualitative study of the rearrangement of the electronic distribution in the molecule, and the aim of this contribution is to demonstrate how in certain cases different topological paradigms are formally connected, with the junction point being the definition of the molecular exciton wave function.

The outcome of the computation of the molecular electronic excited states using a quantum calculation method is, in addition to the transition energy, a series of mathematical objects allowing one to analyze the transition topology. If the reference ground state wave function is, in a given basis (called the canonical basis), written as a Slater determinant, any excited state written based on this ground state wave function is called a single-reference excited state. From this single reference and in a given canonical basis, some methods express excited states as a linear combination of singly excited Slater determinants, which means that the excited state wave function is written as a pondered sum of Slater determinants constructed from the ground state reference, in which one occupied spinorbital (*vide infra*) is replaced by a virtual one. This type of excited state construction is often referred to as a configuration interaction (CI) solely involving singly excited Slater determinants. In our case, the reference ground state wave function can be a Hartree-Fock or a Kohn-Sham Slater determinant, and the excited states calculation methods we deal with in this paper are called configuration interaction singles (CIS), time-dependent Hartree-Fock (TDHF), random-phase approximation (RPA), Tamm-Dancoff approximation (TDA), or time-dependent density functional theory (TDDFT). For more details about the machinery of these methods, see Refs. [22–25]. While in the case of CIS and TDA, the determination of the exciton wave function is very straightforward, for the other methods, it has been subject to the so-called assignment problem which consisted in providing a CI structure to the TDDFT excited state (since the central RPA/TDHF and TDDFT equations have the same structure, the assignment problem is transferable to these methods also) [26, 27].

Based on the outcome of the excited states calculation, one can select an electronic transition of interest and inspect the different hole/particle contributions from the occupied/virtual canonical subspaces for having an insight into the light-induced charge displacement topology. However, in some occurrences, such analyses are quite cumbersome because many of these contributions can be significant while bearing a divergent physical meaning. For the purpose of providing a straightforward picture of the electronic transition topology, multiple tools were developed. Among them, one can cite the detachment/attachment strategy [3, 4, 25, 28–31], which delivers a one-electron charge density function for the hole and for the particle that are generated by photon absorption. This strategy is based on the diagonalization of the so-called difference density matrix (the difference between the excited and ground state density matrices) and a sorting of the resulting "transition occupation numbers" based on their sign. The result of this analysis is a simple identification of the photogenerated depletion and increment zones of charge density. Quantitative insights are then reachable through the manipulation of the detachment/attachment density functions and the definition of quantum metrics [3–5]. On the other hand, one can consider the projection of the exciton wave function in the canonical basis through the so-called transition density matrix [13, 25, 29, 30, 32–43], which singular value decomposition [44] provides the most compact spinorbital representation of the electronic transition. The great advantage of this method is that in most of the cases it condensates the physics of an electronic transition into one couple of hole/particle wave functions.

This chapter first recalls some useful concepts related to the reduced density matrix formalism and its relation to the notion of electron density and density matrix in a canonical space. The detachment/attachment density matrix construction is then exposed in details and is used for quantifying the charge transfer locality through several quantum descriptors. Afterward, the notion of density matrix is extended to electronic transitions through the concept of transition density matrix. The information contained in this particular matrix is shown to be extractable and is discussed in details by introducing the so-called natural transition orbitals. The detachment/attachment and natural transition orbitals formalisms are then compared, and we demonstrate that the difference density matrix is constructed from the direct sum of two matrix products involving only the transition density matrix, that is, the molecular exciton wave function projected into the canonical space (Lemma III.1). It follows that the natural transition orbitals are nothing but the eigenvectors of the detachment/attachment density matrices (Theorem III.1), which is a major conclusion in this contribution since the two formalisms are often introduced as being distinct and belonging to two separate paradigms. This conclusion is finally used for showing that the quantum indices designed for quantifying the charge transfer range and magnitude can be equivalently derived from the detachment/attachment and natural transition orbitals paradigms (Corollary III.1).

All the derivations are performed in the canonical space in the main text, but the important concepts and conclusions are also written in the basis of atomic functions in Appendix B. The calculations performed for this book chapter were done using the G09 software suite [45].

2. Theoretical background

Since this chapter will be mostly dealing with quantum state density matrices, the first paragraph of this section consists in a short reminder about the one-particle reduced density matrices corresponding to single-determinant wave functions.

2.1. One-particle reduced density matrix

We consider an N–electron system, with the N electrons being distributed in L spinorbitals (N occupied, $L-N$ virtual). In this contribution we will write any ground state wave function ψ_0 as an arrangement of the occupied spinorbitals into a single Slater determinant. The density matrix kernel representing the corresponding ground electronic state writes

$$\tilde{\gamma}^0(\mathbf{r}_1,\mathbf{r}_1') = N\sum_{\sigma_1=\alpha,\beta}\int d\mathbf{x}_2\ldots\int d\mathbf{x}_N\,\psi_0(\mathbf{r}_1,\sigma_1,\ldots,\mathbf{x}_N)\psi_0^*(\mathbf{r}_1',\sigma_1,\ldots,\mathbf{x}_N) = \sum_{r=1}^{L}\sum_{s=1}^{L}\varphi_r(\mathbf{r}_1)\left(\gamma^0\right)_{rs}\varphi_s^*(\mathbf{r}_1'), \quad (1)$$

where $\mathbf{x}$ is a four-dimensional variable containing the spatial ($\mathbf{r}$) and the spin-projection (σ) coordinates. The density matrix kernel reduces to the electron density function when $\mathbf{r}_1 = \mathbf{r}_1'$, and its integral over the whole space returns the number of electrons:

$$\tilde{\gamma}^0(\mathbf{r}_1,\mathbf{r}_1) \equiv n_0(\mathbf{r}_1) = \sum_{r=1}^{L}\sum_{s=1}^{L}\varphi_r(\mathbf{r}_1)\left(\gamma^0\right)_{rs}\varphi_s^*(\mathbf{r}_1) \Rightarrow \int_{\mathbb{R}^3} d\mathbf{r}_1\ \tilde{\gamma}^0(\mathbf{r}_1,\mathbf{r}_1) = \int_{\mathbb{R}^3} d\mathbf{r}_1\ n_0(\mathbf{r}_1) = N. \quad (2)$$

The $(\gamma^0)_{rs}$ terms appearing in Eq. (1) are the elements of the one-particle reduced density matrix expressed in the canonical space of spinorbitals $\{\varphi\}$ and can be isolated by integrating the product of $\tilde{\gamma}^0$ with the corresponding spinorbitals

$$(\gamma^0)_{rs} = \int_{\mathbb{R}^3} d\mathbf{r}_1 \int_{\mathbb{R}^3} d\mathbf{r}_1' \ \varphi_r^*(\mathbf{r}_1)\tilde{\gamma}^0(\mathbf{r}_1,\mathbf{r}_1')\varphi_s(\mathbf{r}_1'). \tag{3}$$

Note that generally speaking the $r \times s$ density matrix element in a given spinorbitals space $\{\varphi\}$ for a given quantum state $|\psi\rangle$ writes

$$(\gamma)_{rs} = \left\langle \psi | \hat{r}^{\dagger}\hat{s} | \psi \right\rangle; \ \gamma \in \mathbb{R}^{L\times L} \tag{4}$$

where conventionally r and s indices range from 1 to L. In Eq. (4) we introduced the annihilation and creation operators from the second quantization.

2.2. Detachment and attachment density matrices

One known strategy for formally assigning the depletion and increment zones of charge density appearing upon light absorption is the so-called detachment/attachment formalism. This approach consists in separating the contributions related to light-induced charge removal and accumulation by diagonalizing the one-particle difference density matrix $\gamma^{\Delta} \in \mathbb{R}^{L\times L}$. Such matrix is obtained by taking the difference between the target excited state $|\psi_x\rangle$ and the ground state $|\psi_0\rangle$ density matrices:

$$\gamma^{\Delta} = \gamma^{x} - \gamma^{0}. \tag{5}$$

This density matrix can be projected into the Euclidean space in order to directly visualize the negative and positive contributions to the light-induced charge displacement:

$$n_{\Delta}(\mathbf{r}_1) = \sum_{r=1}^{L}\sum_{s=1}^{L} \varphi_r(\mathbf{r}_1)\left(\gamma^{\Delta}\right)_{rs}\varphi_s^*(\mathbf{r}_1) = n_x(\mathbf{r}_1) - n_0(\mathbf{r}_1). \tag{6}$$

Note that since no fraction of charge has been gained or lost during the electronic transition, the integral of this difference density over all the space is equal to zero:

$$\int_{\mathbb{R}^3} d\mathbf{r}_1 \ n_{\Delta}(\mathbf{r}_1) = \underbrace{\int_{\mathbb{R}^3} d\mathbf{r}_1 \ n_x(\mathbf{r}_1)}_{N} - \overbrace{\int_{\mathbb{R}^3} d\mathbf{r}_1 \ n_0(\mathbf{r}_1)}^{N} = 0. \tag{7}$$

However, visualizing this difference density does not provide a straightforward picture of the transition. The interpretation of the transition in terms of charge density depletion and increment can be made more compact by diagonalizing the difference density matrix:

$$\exists \mathbf{M} \mid \mathbf{M}^{\dagger}\gamma^{\Delta}\mathbf{M} = \mathbf{m} \tag{8}$$

where $\mathbf{m}$ is a diagonal matrix and $\mathbf{M}$ is unitary. Similar to the eigenvalues of a quantum state density matrix, the eigenvalues of γ^{Δ}, contained in $\mathbf{m}$, can be regarded as the occupation

http://dx.doi.org/10.5772/intechopen.70688

numbers of the transition in the canonical space. Those can be negative or positive, corresponding, respectively, to charge removal or accumulation. These eigenvalues can therefore be sorted with respect to their sign:

$$\mathbf{k}_{\pm} = \frac{1}{2}\left(\sqrt{\mathbf{m}^2} \pm \mathbf{m}\right) \tag{9}$$

where $\mathbf{k}_{+}$ (respectively, $\mathbf{k}_{-}$) is a diagonal matrix storing the positive (absolute value of negative) eigenvalues of the difference density matrix. These two diagonal matrices can be separately backtransformed to provide the so-called detachment (*d*) and attachment (*a*) density matrices and the corresponding charge densities:

$$\mathbf{Mk}_{-}\mathbf{M}^{\dagger} = \gamma^{d} \overset{\mathbb{R}^3}{\rightarrow} n_d(\mathbf{r}) = \sum_{r=1}^{L}\sum_{s=1}^{L}\left(\gamma^{d}\right)_{rs} \varphi_r(\mathbf{r})\varphi_s^{*}(\mathbf{r});\ \mathbf{Mk}_{+}\mathbf{M}^{\dagger} = \gamma^{a} \overset{\mathbb{R}^3}{\rightarrow} n_a(\mathbf{r}) = \sum_{r=1}^{L}\sum_{s=1}^{L}\left(\gamma^{a}\right)_{rs} \varphi_r(\mathbf{r})\varphi_s^{*}(\mathbf{r}). \tag{10}$$

These detachment/attachment densities ($n_d(\mathbf{r})$ and $n_a(\mathbf{r})$) are then nothing but the hole and particle densities we were seeking. These densities are reproduced in **Figure 1** for two paradigmatic cases of electronic transitions: one local transition and one long-range charge transfer. In the next paragraph, we will see how the locality of a charge transfer can be quantified using the detachment/attachment charge densities.

2.3. Quantifying the charge transfer locality

One possible strategy for evaluating the magnitude of the electronic structure reorganization is to compute the spatial overlap between the hole and the particle. This is possible through the assessment of a normalized, dimensionless quantity named ϕ_S:

$$\phi_S = \vartheta_x^{-1}\int_{\mathbb{R}^3} d\mathbf{r}\sqrt{n_d(\mathbf{r})n_a(\mathbf{r})} \in [0;1];\ \vartheta_x = \frac{1}{2}\sum_{q=d,\,a}\int_{\mathbb{R}^3} d\mathbf{r}\, n_q(\mathbf{r}) \tag{11}$$

where ϑ_x is a normalization factor (the integral of detachment/attachment density over all the space). Obviously, a long-range charge transfer means a low hole/particle overlap and will correspond to a low value for ϕ_S. Conversely, a local transition will be characterized by a

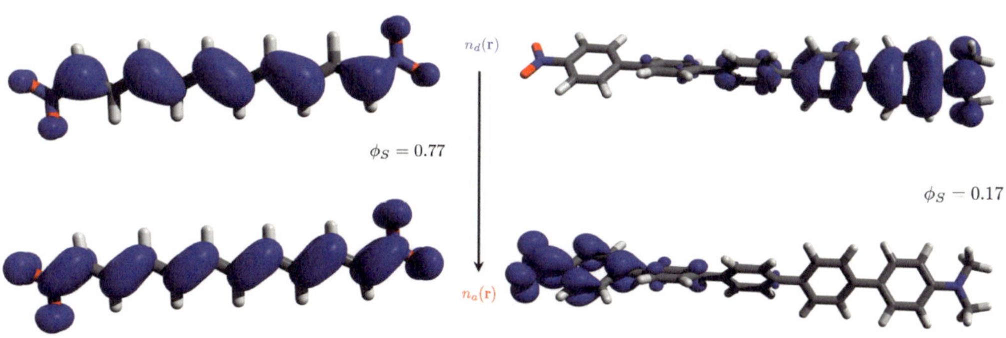

Figure 1. Illustration of a local (left) and long-range (right) transition using detachment/attachment densities and the ϕ_S index.

higher ϕ_S value. This is clearly illustrated in **Figure 1** where the ϕ_S value drops from 0.77 to 0.17 when going from an electronic transition exhibiting a large hole/particle overlap to a long-range charge transfer. These two cases are used solely to illustrate the potentiality of the ϕ_S quantum metric to assess the locality of a charge transfer. The computation of ϕ_S is schematically pictured in the top of **Figure 2**.

It has also been demonstrated that ϕ_S can be used for performing a diagnosis on the exchange-correlation functional used for computing the transition energy within the framework of TDDFT [3].

An additional quantitative strategy consists in computing the charge effectively displaced during the transition. The difference between the hole/particle and the effectively displaced charge density is illustrated in **Figure 2**: since there can be some overlap between the hole and the particle densities, the global outcome (the "bilan") of the transition in terms of charge displacement is not the detachment and attachment but the negative and positive contributions to the difference density, which can be obtained by taking the difference between the attachment and detachment charge densities at every point of space. Indeed, from

$$\mathbf{m} = \mathbf{k}_+ - \mathbf{k}_- \Rightarrow \gamma^\Delta = \mathbf{MmM}^\dagger = \mathbf{Mk}_+\mathbf{M}^\dagger - \mathbf{Mk}_-\mathbf{M}^\dagger = \gamma^a - \gamma^d \tag{12}$$

we can write

$$n_\Delta(\mathbf{r}) = n_a(\mathbf{r}) - n_d(\mathbf{r}), \tag{13}$$

and introduce the *actual* displacement charge density functions

$$n_\pm(\mathbf{r}) = \frac{1}{2}\left\{\sqrt{n_\Delta^2(\mathbf{r})} \pm n_\Delta(\mathbf{r})\right\} \tag{14}$$

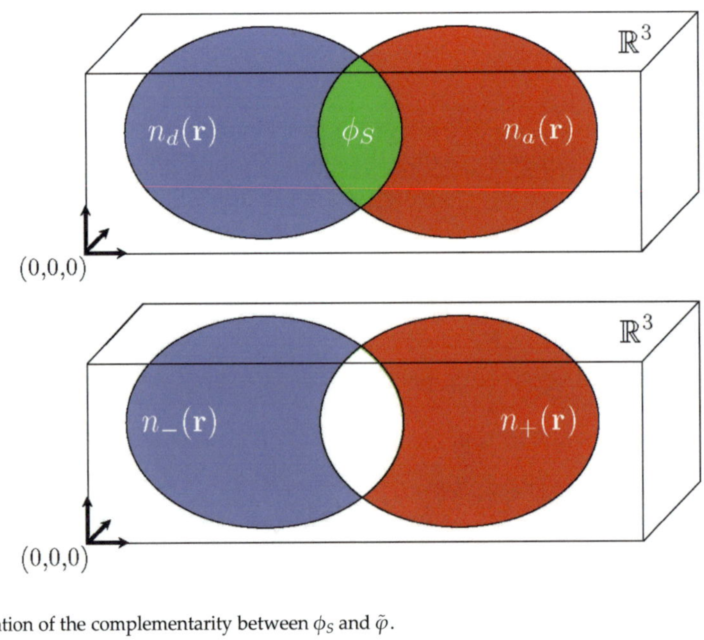

Figure 2. Illustration of the complementarity between ϕ_S and $\tilde{\varphi}$.

so the splitting operation is performed based on the sign of function entries in the three dimensions of space instead of transition occupation numbers. From this separation we can compute the normalized displaced charge:

$$\frac{\vartheta_x^{-1}}{2}\sum_{s=+,-}\int_{\mathbb{R}^3} d\mathbf{r}\ n_s(\mathbf{r}) = \tilde{\varphi} \in [0;1]. \tag{15}$$

Obviously, splitting the transition occupation numbers and computing the detachment/attachment overlap are complementary to the integration of the negative and positive contributions to the difference density function: the ϕ_S descriptor provides an information related to the locality of the charge transfer, while the $\tilde{\varphi}$ metric relates the amount of charge transferred during the transition.

These two complementary approaches have been associated into a final, general quantum metric of charge transfer:

$$\psi = 2\pi^{-1}\underbrace{\arctan\left(\frac{\phi_S}{\tilde{\varphi}}\right)}_{\theta_S} = \frac{2\theta_S}{\pi} \in [0;1[\tag{16}$$

which, as it was the case for ϕ_S and $\tilde{\varphi}$, is normalized and dimensionless. The ψ metric can be interpreted as the normalized angle resulting from the joint projection of ϕ_S and $\tilde{\varphi}$ in a complex plane ($\tilde{\varphi}$ being along the real axis and ϕ_S the imaginary one). Such projection is characterized by a θ_S angle with the real axis (see Ref. [5]), taking values ranging from 0 to $\pi/2$. Therefore, the $2\pi^{-1}$ factor in Eq. (16) is there to ensure that ψ is normalized. Note that there exists multiple ways to derive the three quantum metrics exposed in this paragraph, as mentioned in Ref. [5].

Figure 3 represents the ψ projection for a series of dyes. These chromophores are constituted by an electron-donor fragment conjugated to an acceptor moiety through a molecular bridge with a variable size (i.e., a variable number of subunits).

We see that when the first excited state of these dyes is computed using TDDFT with the hybrid PBE0 exchange-correlation functional [46, 47] and a triple-zeta split-valence Gaussian basis set with diffuse and polarization functions on every atom [48], increasing the number of bridge subunits leads to a net decrease in the ψ projection angle. It is therefore very clear from **Figure 3** that increasing the length of the bridge for this family of dyes leads to an increase of the charge transfer character of the first transition, when computed at the above-mentioned level of theory.

The following paragraph details another known strategy providing a straightforward qualitative analysis of the charge transfer topology, based on another type of density matrix: the transition density matrix.

2.4. Transition density matrix and natural transition orbitals

In the following section we will be interested in the determination of the exciton wave function and its use for providing the most compact representation of an electronic transition. More

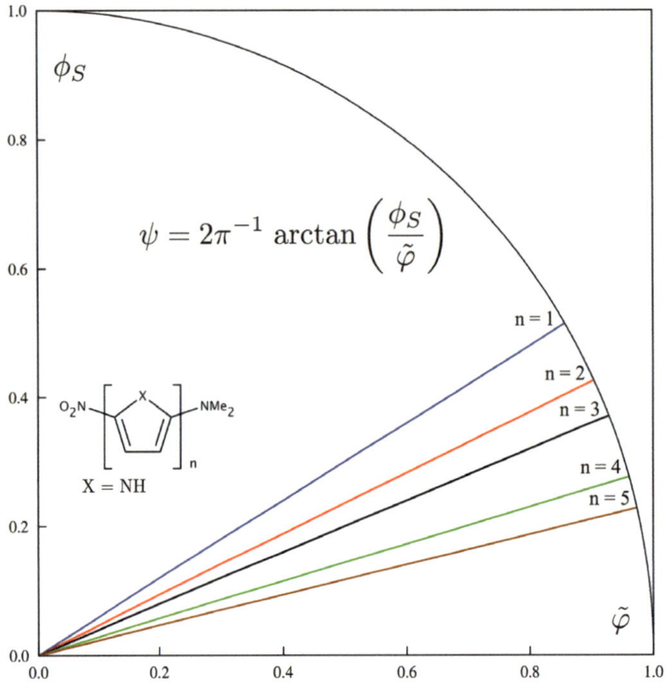

Figure 3. Illustration of the evolution of the ψ index value for the first excited state of a series of push-pull dyes, computed at the PBE0/6-311++G(2d,p)//PBE0/6-311G(d,p) level of theory in vacuum.

particularly, this paragraph exposes how we can find an alternative basis to the canonical one and reduce the picture of the transition to one couple of hole/particle wave functions. The following formalism is applied to the case of quantum excited states that can be written as a linear combination of singly excited Slater determinants, constructed from the single-reference wave function (ψ_0) where the spinorbital φ_i from the occupied canonical subspace has been replaced by the φ_a spinorbital belonging to the virtual canonical subspace. In these conditions, the xth excited electronic state writes

$$|\psi_x\rangle = \sum_{i=1}^{N} \sum_{a=N+1}^{L} z_x^{-1/2} \left(\gamma^{0x}\right)_{ia} |\psi_i^a\rangle; \; |\psi_i^a\rangle = \hat{a}^{\dagger}\hat{i}|\psi_0\rangle \tag{17}$$

where again we introduced the annihilation $\hat{i}$ and creation $\hat{a}^{\dagger}$ operators from the second quantization, so we actually see that $|\psi_i^a\rangle$ is obtained by annihilating the electron in the ith spinorbital from the ground state wave function and creating an electron in the ath one. In Eq. (17),

$$z_x = \mathrm{tr}\left(\gamma^{0x}\gamma^{0x\dagger}\right) = \mathrm{tr}\left(\gamma^{0x\dagger}\gamma^{0x}\right) \tag{18}$$

is a normalization factor and $\left(\gamma^{0x}\right)_{ia}$ is a transition density matrix element for the $0 \rightarrow x$ state transition. Transition density matrix elements can be extracted from the exciton wave function:

http://dx.doi.org/10.5772/intechopen.70688

$$\begin{aligned}\tilde{\gamma}^{0x}(\mathbf{r}_1,\mathbf{r}_1') &= N\sum_{\sigma_1=\alpha,\beta}\int d\mathbf{x}_2\ldots\int d\mathbf{x}_N\,\psi_0(\mathbf{r}_1,\ldots,\mathbf{x}_N)\psi_x^*(\mathbf{r}_1',\ldots,\mathbf{x}_N)\\ &= z_x^{-1/2}\sum_{i=1}^{N}\sum_{a=N+1}^{L}\varphi_i(\mathbf{r}_1)\left(\gamma^{0x}\right)_{ia}\varphi_a^*(\mathbf{r}_1'),\end{aligned} \tag{19}$$

That is, the so-called transition density matrix kernel locating the hole ($\mathbf{r}_1$) in the ground state and the particle ($\mathbf{r}_1'$) in the excited state. Similarly to the one-particle reduced density matrix in Eq. (4), the transition density matrix elements write

$$\begin{aligned}z_x^{-1/2}\left(\gamma^{0x}\right)_{ia} &= \left\langle\psi_0\middle|\hat{i}^{\dagger}\hat{a}\middle|\psi_x\right\rangle = \sum_{j=1}^{N}\sum_{b=N+1}^{L} z_x^{-1/2}\left(\gamma^{0x}\right)_{jb}\left\langle\psi_0\middle|\hat{i}^{\dagger}\hat{a}\middle|\psi_j^b\right\rangle\\ &= \sum_{j=1}^{N}\sum_{b=N+1}^{L} z_x^{-1/2}\left(\gamma^{0x}\right)_{jb}\overbrace{\left\langle\psi_i^a\middle|\psi_j^b\right\rangle}^{\delta_{ij}\delta_{ab}} = z_x^{-1/2}\left(\gamma^{0x}\right)_{ia}.\end{aligned} \tag{20}$$

Note that we conventionally set the i,j and a,b indices to match spinorbitals, respectively, belonging exclusively to the occupied and virtual canonical subspaces, while r and s indices have no restricted attribution to a given subspace. Similarly to the quantum state electron density function, one can deduce the expression of the one-particle transition density from the transition density matrix kernel:

$$\begin{aligned}n^{0x}(\mathbf{r}_1) &= z_x^{-1/2}\sum_{i=1}^{N}\sum_{a=N+1}^{L}\varphi_i(\mathbf{r}_1)\left(\gamma^{0x}\right)_{ia}\varphi_a^*(\mathbf{r}_1)\\ \Rightarrow \int_{\mathbb{R}^3} d\mathbf{r}_1\; n^{0x}(\mathbf{r}_1) &= z_x^{-1/2}\sum_{i=1}^{N}\sum_{a=N+1}^{L}\left(\gamma^{0x}\right)_{ia}\underbrace{\langle\varphi_a|\varphi_i\rangle}_{\delta_{ia}} = 0\end{aligned} \tag{21}$$

where the δ_{ia} Kronecker delta is systematically vanishing since φ_i and φ_a spinorbitals never belong to the same subspace. Here again, we will take advantage of the possibility to use finite mathematical objects such as matrices and perform a reduction of the one-particle transition density matrix size: since we know that i and a indices are restricted to occupied and virtual subspaces, we can introduce the normalized transition density matrix $\mathbf{T}$ (that we will call transition density matrix in the following):

$$z_x^{-1/2}\left(\gamma^{0x}\right)_{ia} \leftrightarrow (\mathbf{T})_{ic} \quad (c = a - N) \tag{22}$$

so the connection between the two matrices is trivial:

$$z_x^{-1/2}\gamma^{0x} = \begin{pmatrix} 0_{N\times N} & \mathbf{T}\\ 0_{(L-N)\times N} & 0_{(L-N)\times(L-N)}\end{pmatrix};\; z_x^{-1/2}\gamma^{0x}\in\mathbb{R}^{L\times L}\leftrightarrow \mathbf{T}\in\mathbb{R}^{N\times(L-N)} \tag{23}$$

where $0_{k\times l}$ refers to the zero matrix with $k\times l$ dimensions. For the sake of simplicity, we will use 0_o and 0_v for the occupied × occupied and virtual × virtual zero blocks and $0_{o\times v}$ and $0_{v\times o}$ for the out-diagonal blocks.

We will now focus on **T**. This matrix contains the information related to the transition we seek, and similarly to the difference density matrix, we will extract this information by diagonalizing **T**. However, since **T** is not square but rectangular (we rarely have the same number of occupied and virtual orbitals), the diagonalization process is named singular value decomposition (SVD) [44] and takes the form

$$\exists \mathbf{O}, \mathbf{V} \mid \mathbf{O}^{\dagger}\mathbf{T}\mathbf{V} = \boldsymbol{\lambda}. \tag{24}$$

The diagonal $\boldsymbol{\lambda}$ entries are called the singular values of **T**. Due to the dimensions of $\boldsymbol{\lambda}$, the number of singular values is equal to the dimensions of the lowest subspace (i.e., N or $L-N$). Most often, the number of virtual orbitals is larger than the number of occupied orbitals. Therefore, from now on we will assume that $N<L-N$.

While from the diagonalization of γ^{Δ} we could build detachment/attachment densities, here we will use the left and right eigenvectors of **T** for rotating the occupied and virtual canonical subspaces into the so-called occupied/virtual natural transition orbital (NTO) spaces:

$$\varphi_i^o(\mathbf{r}) = \sum_{j=1}^{N} (\mathbf{O})_{ji}\varphi_j(\mathbf{r}) \overset{(\boldsymbol{\lambda})_{ii}}{\leftrightarrow} \varphi_i^v(\mathbf{r}) = \sum_{j=1}^{L-N} (\mathbf{V})_{ji}\varphi_{N+j}(\mathbf{r}), \tag{25}$$

where i ranges from 1 to N. We have built N couples of occupied/virtual NTOs, each couple being characterized by the corresponding singular value $(\boldsymbol{\lambda})_{ii}$. The great advantage of performing an SVD on **T** is that in most of the cases, only one singular value is predominant, which means that we can condensate all the physics of an electronic transition into one couple of occupied/virtual NTOs, as represented in **Figure 4**.

We can conclude that, similarly to the usual quantum state natural orbitals which constitute the basis in which the quantum state density matrix is diagonal, the NTOs provide the most compact representation of the electronic transition and can be used to rewrite the expression of the electronic excited state and the transition density matrix kernel (the exciton wave function):

$$|\psi_x\rangle = \sum_{i=1}^{N} (\boldsymbol{\lambda})_{ii}|\psi_{o,i}^{v,i}\rangle = \sum_{i=1}^{N} (\boldsymbol{\lambda})_{ii}\hat{q}_i^{v\dagger}\hat{q}_i^{o}|\psi_0\rangle;\ \tilde{\gamma}^{0x}(\mathbf{r}_1,\mathbf{r}_1') = \sum_{i=1}^{N} (\boldsymbol{\lambda})_{ii}\varphi_i^o(\mathbf{r}_1)\varphi_i^{v*}(\mathbf{r}_1') \tag{26}$$

where this time the creation/annihilation operators are bearing the "o" and "v" superscripts, reminding that we are annihilating an electron in the ith occupied (o) NTO and creating one electron in the ith virtual (v) NTO. Since we know that usually one singular value is predominant, we can clearly identify the hole and particle wave functions and state, upon light absorption, from where the electron goes and where it arrives.

Multiplying **T** by its own transpose and vice versa leads to two square matrices with interesting properties:

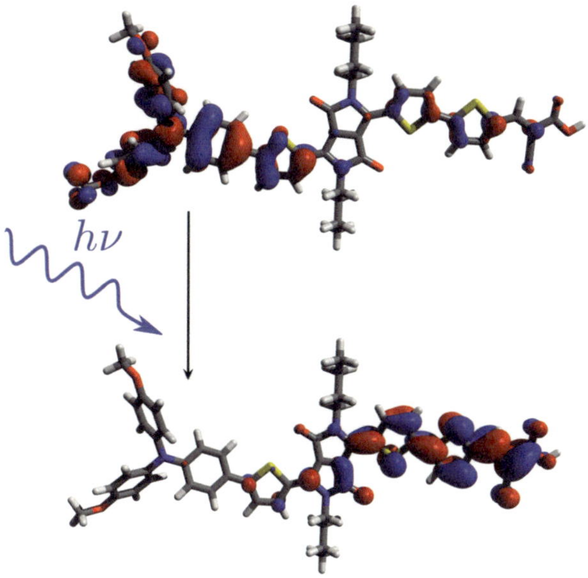

Figure 4. Illustration of the hole (top) and particle (bottom) wave functions, that is, the predominant couple of occupied (top) and virtual (bottom) NTOs for a random push-pull chromophore experiencing a photoinduced charge transfer.

$$\mathbf{TT}^{\dagger} \in \mathbb{R}^{N\times N}; \ \mathbf{T}^{\dagger}\mathbf{T} \in \mathbb{R}^{(L-N)\times(L-N)}. \tag{27}$$

Due to their structure, these two new matrices share the same eigenvectors than $\mathbf{T}$

$$\mathbf{O}^{\dagger}\mathbf{TT}^{\dagger}\mathbf{O} = \boldsymbol{\lambda}_o^2; \ \mathbf{V}^{\dagger}\mathbf{T}^{\dagger}\mathbf{TV} = \boldsymbol{\lambda}_v^2 \tag{28}$$

with, considering $N<L-N$, the following rules for their eigenvalues:

$$(\boldsymbol{\lambda})_{ii}^2 = \left(\boldsymbol{\lambda}_o^2\right)_{ii} = \left(\boldsymbol{\lambda}_v^2\right)_{ii} \ \forall i \leq N; \quad \boldsymbol{\lambda}_v^2 = \boldsymbol{\lambda}_o^2 \oplus 0_v. \tag{29}$$

These rules can be demonstrated by developing the product of λ with its own transpose:

$$\boldsymbol{\lambda\lambda}^{\dagger} = \mathbf{O}^{\dagger}\mathbf{T}\underbrace{\mathbf{VV}^{\dagger}}_{I_v}\mathbf{T}^{\dagger}\mathbf{O} = \mathbf{O}^{\dagger}\mathbf{TT}^{\dagger}\mathbf{O} \tag{30}$$

where I_v is the $(L-N)\times(L-N)$ identity matrix. Due to the dimensions of $\boldsymbol{\lambda}$ and its diagonal structure, we can write

$$\boldsymbol{\lambda} \in \mathbb{R}^{N\times(L-N)} \ \Rightarrow \ \boldsymbol{\lambda\lambda}^{\dagger} \in \mathbb{R}^{N\times N} \quad ; \quad (\boldsymbol{\lambda})_{ij} = 0 \ \forall i \neq j \ \Rightarrow \ \boldsymbol{\lambda\lambda}^{\dagger} = \boldsymbol{\lambda}_o^2. \tag{31}$$

Similarly, we have for $\boldsymbol{\lambda}^{\dagger}\boldsymbol{\lambda}$

$$\boldsymbol{\lambda}^{\dagger}\boldsymbol{\lambda} = \mathbf{V}^{\dagger}\mathbf{T}^{\dagger}\underbrace{\mathbf{OO}^{\dagger}}_{I_o}\mathbf{TV} = \mathbf{V}^{\dagger}\mathbf{T}^{\dagger}\mathbf{TV} \tag{32}$$

and

$$\boldsymbol{\lambda} \in \mathbb{R}^{N\times(L-N)} \Rightarrow \boldsymbol{\lambda}^{\dagger}\boldsymbol{\lambda} \in \mathbb{R}^{(L-N)\times(L-N)} \quad ; \quad (\boldsymbol{\lambda})_{ij} = 0 \ \forall i \neq j \Rightarrow \boldsymbol{\lambda}^{\dagger}\boldsymbol{\lambda} = \boldsymbol{\lambda}_v^2. \tag{33}$$

Multiplying Eq. (28) by the left by $\mathbf{T}^{\dagger}\mathbf{O}$ or $\mathbf{TV}$ leads to two new eigenvalue problems:

$$\mathbf{T}^{\dagger}\mathbf{O}\left(\mathbf{O}^{\dagger}\mathbf{T}\mathbf{T}^{\dagger}\mathbf{O} = \lambda_o^2\right) \Leftrightarrow \mathbf{T}^{\dagger}\mathbf{T}\underbrace{\mathbf{T}^{\dagger}\mathbf{O}}_{\mathbf{V}_o} = \mathbf{T}^{\dagger}\mathbf{O}\lambda_o^2 \quad ; \quad \mathbf{TV}\left(\mathbf{V}^{\dagger}\mathbf{T}^{\dagger}\mathbf{TV} \Leftrightarrow \lambda_v^2\right) = \mathbf{T}\mathbf{T}^{\dagger}\underbrace{\mathbf{TV}}_{\mathbf{O}_v} = \mathbf{TV}\lambda_v^2 \tag{34}$$

where $\mathbf{V}_o \in \mathbb{R}^{(L-N)\times N}$ contains the N eigenvectors of $\mathbf{T}^{\dagger}\mathbf{T}$ with a nonvanishing eigenvalue (i.e., the N first columns of $\mathbf{V}$) and $\mathbf{O}_v \in \mathbb{R}^{N\times(L-N)}$ is the juxtaposition of $\mathbf{O}$ and $L-2N$ zero columns. The results in Eq. (34) prove that the eigenvectors of each of the two matrices in Eq. (27) can be found from the eigenvectors of the other one and that both matrices share the same nonvanishing eigenvalues, as mentioned in Eq. (29).

3. Bridging the detachment/attachment and NTO paradigms

We now have two general strategies for qualitatively studying the topology of the light-induced electronic cloud polarization, and the locality of this electronic structure reorganization can be quantified. This section is devoted to single-reference excited states calculation methods that express the electronic excited state as a linear combination of singly excited Slater determinants and brings the rigorous demonstration that in such case, the three quantum metrics we previously designed can be formally equivalently derived from the difference density matrix or the transition density matrix. This result is the corollary to a theorem stating that the occupied/virtual NTOs are nothing but the eigenvectors of the detachment/attachment density matrices.

3.1. Expression of the quantum state density matrices in the canonical space

In this paragraph we elucidate the structure of the difference density matrix by developing the full expression of the excited state density matrix in the canonical space.

Lemma III.1 *The difference density matrix is the direct sum of* $-\mathbf{T}\mathbf{T}^{\dagger}$ and $\mathbf{T}^{\dagger}\mathbf{T}$.

Proof. We start by writing the expression of the ground state density matrix: from Eq. (4) it follows that for an N–electron single-determinant ground state wave function,

$$\forall(r,s) \mid r \leq N \text{ and } s \leq N, \ (\gamma)_{rs} = \delta_{rs} \quad ; \quad \forall(r,s) \mid r > N \text{ and/or } s > N, \ (\gamma)_{rs} = 0. \tag{35}$$

It follows that the ground state one-particle density matrix in the canonical space writes

$$\gamma^0 = I_o \oplus 0_v. \tag{36}$$

If now we rewrite the electronic excited state $|\psi_x\rangle$ from Eq. (17) using the normalized transition density matrix elements, we have

$$|\psi_x\rangle = \sum_{i=1}^{N} \sum_{a=N+1}^{L} (\mathbf{T})_{ic} |\psi_i^a\rangle \quad (c = a - N). \tag{37}$$

From now on we will operate a systematic index shift between matrix elements and virtual orbitals implied in the singly excited Slater determinants. Since the excited state wave function is normalized, we can write

$$\begin{aligned} 1 = \langle \psi_x | \psi_x \rangle &= \sum_{i,j=1a,\, b=N+1}^{N \quad L} (\mathbf{T})_{jd}^* (\mathbf{T})_{ic} \overbrace{\left\langle \psi_j^b | \psi_i^a \right\rangle}^{\delta_{ij}\delta_{ab}} \qquad (d = b - N) \\ &= \sum_{i=1}^{N} \sum_{a=N+1}^{L} (\mathbf{T})_{ic}^* (\mathbf{T})_{ic} = \sum_{i=1}^{N} \sum_{a=N+1}^{L} (\mathbf{T})_{ic} (\mathbf{T}^\dagger)_{ci} = \begin{cases} \mathrm{tr}(\mathbf{T}\mathbf{T}^\dagger) \\ \mathrm{tr}(\mathbf{T}^\dagger \mathbf{T}) \end{cases} \end{aligned} \tag{38}$$

and, since the trace of a matrix is an unitary invariant,

$$\mathrm{tr}(\boldsymbol{\lambda}\boldsymbol{\lambda}^\dagger) = \mathrm{tr}(\boldsymbol{\lambda}^\dagger \boldsymbol{\lambda}) = 1. \tag{39}$$

Using the second quantization, we might rewrite $|\psi_x\rangle$

$$|\psi_x\rangle = \sum_{i=1}^{N} \sum_{a=N+1}^{L} (\mathbf{T})_{ic} \; \widehat{a}^{\dagger} \widehat{i} |\psi_0\rangle \tag{40}$$

and the $r \times s$ density matrix element for the xth excited state writes

$$\begin{aligned} (\gamma^x)_{rs} = \left\langle \psi_x | \widehat{r}^{\dagger} \widehat{s} | \psi_x \right\rangle &= \sum_{i,j=1a,\, b=N+1}^{N \quad L} (\mathbf{T})_{jd}^* (\mathbf{T})_{ic} \left\langle \psi_j^b | \widehat{r}^{\dagger} \widehat{s} | \psi_i^a \right\rangle \\ &= \sum_{i,j=1a,\, b=N+1}^{N \quad L} (\mathbf{T})_{jd}^* (\mathbf{T})_{ic} \left\langle \psi_0 | \widehat{j}^{\dagger} \widehat{b} \widehat{r}^{\dagger} \widehat{s} \widehat{a}^{\dagger} \widehat{i} | \psi_0 \right\rangle. \end{aligned} \tag{41}$$

We will now apply Wick's theorem to the expression of the excited state density matrix written using our fermionic second quantization operators. According to this theorem, one can rewrite Eq. (41) as a combination of products of expectation values of couples of the second quantization operators implied in the expression of γ^x. Since we are working with fermionic operators, a phase is assigned to each term of this sum with the form $(-1)^{\varrho_l}$ where l corresponds to the position of the term in the sum. Note that a number is also assigned to the position of each fermionic operator both in the original expression of γ^x and after expanding it into a sum of terms. **Figure 5** illustrates the case of γ^x, which can be decomposed into a sum of three nonvanishing terms. The central part of the figure shows how each term is constructed by associating a creation to an annihilation operator. Note that other operator pairings are possible, but their expectation value is vanishing due to the fact that the associated operators do not belong to the same subspace (occupied or virtual). The right part of **Figure 5** shows how the label sequence of the operators has been rearranged for each term.

$$\underset{1\,2\,3\,4\,5\,6}{\langle\psi_0|\hat{j}^\dagger\hat{b}\hat{r}^\dagger\hat{s}\hat{a}^\dagger\hat{i}|\psi_0\rangle} = \left\{\begin{array}{c} \underset{1\,2\,3\,4\,5\,6}{\langle\psi_0|\hat{j}^\dagger\hat{b}\hat{r}^\dagger\hat{s}\hat{a}^\dagger\hat{i}|\psi_0\rangle} \\ \underset{1\,2\,3\,4\,5\,6}{\langle\psi_0|\hat{j}^\dagger\hat{b}\hat{r}^\dagger\hat{s}\hat{a}^\dagger\hat{i}|\psi_0\rangle} \\ \underset{1\,2\,3\,4\,5\,6}{\langle\psi_0|\hat{j}^\dagger\hat{b}\hat{r}^\dagger\hat{s}\hat{a}^\dagger\hat{i}|\psi_0\rangle} \end{array}\right\} = \begin{array}{l} (-1)^{\varrho_1}\underset{1\,6}{\langle\psi_0|\hat{j}^\dagger\hat{i}|\psi_0\rangle}\underset{2\,5}{\langle\psi_0|\hat{b}\hat{a}^\dagger|\psi_0\rangle}\underset{3\,4}{\langle\psi_0|\hat{r}^\dagger\hat{s}|\psi_0\rangle} \\ +(-1)^{\varrho_2}\underset{1\,4}{\langle\psi_0|\hat{j}^\dagger\hat{s}|\psi_0\rangle}\underset{2\,5}{\langle\psi_0|\hat{b}\hat{a}^\dagger|\psi_0\rangle}\underset{3\,6}{\langle\psi_0|\hat{r}^\dagger\hat{i}|\psi_0\rangle} \\ +(-1)^{\varrho_3}\underset{1\,6}{\langle\psi_0|\hat{j}^\dagger\hat{i}|\psi_0\rangle}\underset{4\,5}{\langle\psi_0|\hat{s}\hat{a}^\dagger|\psi_0\rangle}\underset{2\,3}{\langle\psi_0|\hat{b}\hat{r}^\dagger|\psi_0\rangle} \end{array}$$

Figure 5. Wick's theorem applied to single-reference excited state density matrices.

Once the excited state density matrix is developed, one can write a bijection $f_l(x)=y$ between the original sequence of operators label (here 1, …, 6) and the one characterizing each term ($l=1,2,3$). The ϱ_l value is then obtained by counting the number of pairs of projections satisfying

$$(x_1, x_2) \mid \{x_1 < x_2;\ f_l(x_1) > f_l(x_2)\} \tag{42}$$

in the bijection. For example, for the first term ($l=1$), the ($x_1=2$, $x_2=5$) pair satisfies this condition, because $f_1(2)=6>3=f_1(5)$. The evaluation of the phase to be assigned to the first term ($l=1$) reported in **Figure 5** is fully detailed in **Figure 6**. The deduction of the phase for $l=2$ and 3 is given in Appendix (**Figures 7** and **8**).

For each term in the developed expression of γ^x, six permutations of its factors are possible without affecting the phase, for the parity of ϱ_l is guided only by the primary association of creation/annihilation operators characterizing the lth term. According to what precedes, we are now able to write the $r \times s$ excited state density matrix element:

$$(\gamma^x)_{rs} = \sum_{i,j=1}^{N}\sum_{a,b=N+1}^{L}(\mathbf{T})^*_{jd}(\mathbf{T})_{ic}(\mathcal{I}_1 - \mathcal{I}_2 + \mathcal{I}_3) \tag{43}$$

with

$$\mathcal{I}_1 = \overbrace{\left\langle\psi_0|\hat{j}^\dagger\hat{i}|\psi_0\right\rangle}^{\delta_{ij}}\underbrace{\left\langle\psi_0|\hat{b}\hat{a}^\dagger|\psi_0\right\rangle}_{\delta_{ab}}\overbrace{\left\langle\psi_0|\hat{r}^\dagger\hat{s}|\psi_0\right\rangle}^{\delta_{rs}n_r} \tag{44}$$

where n_r is the occupation number of spinorbital r (see Eq. (35) for more details). For $\mathcal{I}_2$ we have

http://dx.doi.org/10.5772/intechopen.70688

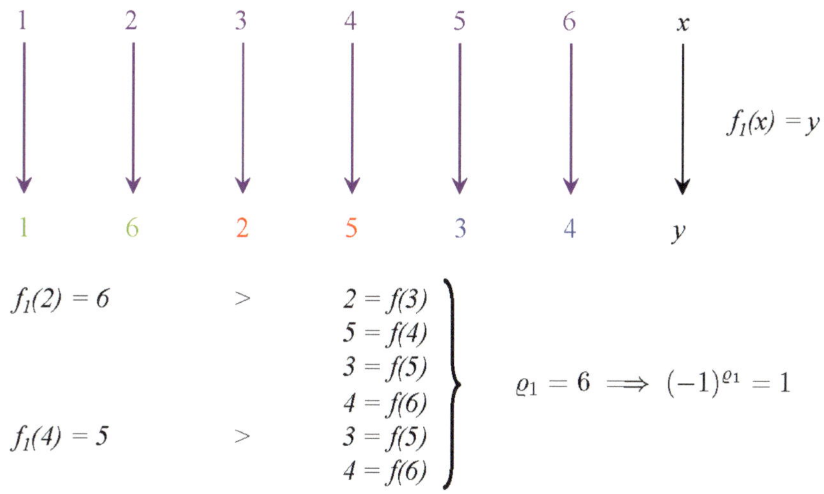

Figure 6. Illustration of the evaluation of ϱ_1.

$$\mathcal{I}_2 = \overbrace{\left\langle \psi_0 | \hat{j}^{\dagger} \hat{s} | \psi_0 \right\rangle}^{\delta_{js}} \underbrace{\left\langle \psi_0 | \hat{b} \hat{a}^{\dagger} | \psi_0 \right\rangle}_{\delta_{ab}} \overbrace{\left\langle \psi_0 | \hat{r}^{\dagger} \hat{i} | \psi_0 \right\rangle}^{\delta_{ri}} \tag{45}$$

and for $\mathcal{I}_3$,

$$\mathcal{I}_3 = \overbrace{\left\langle \psi_0 | \hat{j}^{\dagger} \hat{i} | \psi_0 \right\rangle}^{\delta_{ij}} \underbrace{\left\langle \psi_0 | \hat{s} \hat{a}^{\dagger} | \psi_0 \right\rangle}_{\delta_{sa}} \overbrace{\left\langle \psi_0 | \hat{b} \hat{r}^{\dagger} | \psi_0 \right\rangle}^{\delta_{br}}. \tag{46}$$

Note that since i and j are corresponding to occupied spinorbitals, writing δ_{js} is equivalent to writing $\delta_{js} n_s$ and is not vanishing only when φ_s belongs to the occupied subspace. This is also the case for δ_{ri}. On the other hand, since φ_a and φ_b belong to the virtual subspace, writing δ_{sa} is equivalent to writing $\delta_{sa}(1 - n_s)$ and is not vanishing only when s is superior to N. Note also that writing δ_{ab} when dealing with spinorbitals corresponds to δ_{cd} when working with matrix elements (see Eqs. (37) and (38)). Therefore, $(\gamma^x)_{rs}$ now writes

$$\begin{aligned} \left(\gamma^x\right)_{rs} &= \delta_{rs} n_r \overbrace{\left[\sum_{i,j=1}^{N} \sum_{a,b=N+1}^{L} (\mathbf{T})^*_{jd} (\mathbf{T})_{ic} \delta_{ij} \delta_{ab} \right]}^{\mathrm{tr}\left(\mathbf{T}\mathbf{T}^{\dagger}\right) = \mathrm{tr}\left(\mathbf{T}^{\dagger}\mathbf{T}\right) = 1} \\ &- \sum_{i,j=1}^{N} \sum_{a,b=N+1}^{L} (\mathbf{T})^*_{jd} (\mathbf{T})_{ic} \delta_{ab} \delta_{js} n_s \delta_{ri} n_r \\ &+ \sum_{i,j=1}^{N} \sum_{a,b=N+1}^{L} (\mathbf{T})^*_{jd} (\mathbf{T})_{ic} \delta_{ij} \delta_{br} (1 - n_r) \delta_{sa} (1 - n_s), \end{aligned} \tag{47}$$

that is,

$$\left(\gamma^{x}\right)_{rs}=\delta_{rs}n_r-\left(\mathbf{T}\mathbf{T}^{\dagger}\right)_{ij}\delta_{js}n_s\delta_{ri}n_r+\left(\mathbf{T}^{\dagger}\mathbf{T}\right)_{dc}(1-n_r)\delta_{d(r-N)}(1-n_s)\delta_{c(s-N)}. \tag{48}$$

We see that the first and second terms belong to the occupied × occupied block, while the third term belongs to the virtual × virtual one. According to this, the excited state density matrix in the canonical space finally writes

$$\gamma^{x}=\left(I_o-\mathbf{T}\mathbf{T}^{\dagger}\right)\oplus\mathbf{T}^{\dagger}\mathbf{T}. \tag{49}$$

Subtracting the ground state density matrix taken from Eq. (36) to γ^{x} gives γ^{Δ}

$$\gamma^{\Delta}=-\mathbf{T}\mathbf{T}^{\dagger}\oplus\mathbf{T}^{\dagger}\mathbf{T}.\quad\blacksquare \tag{50}$$

Since $\mathbf{T}\mathbf{T}^{\dagger}$ and $\mathbf{T}^{\dagger}\mathbf{T}$ have positive eigenvalues (i.e., they are positive definite), we deduce

$$(\mathbf{m})_{ii}\leq 0\ \ \forall i\leq N;\ (\mathbf{m})_{aa}\geq 0\ \ \forall a>N. \tag{51}$$

Therefore, we must have that

$$\mathbf{T}\mathbf{T}^{\dagger}\oplus 0_v=\gamma^{d}\quad;\quad 0_o\oplus\mathbf{T}^{\dagger}\mathbf{T}=\gamma^{a} \tag{52}$$

which obviously leads to

$$\gamma^{\Delta}=\gamma^{a}-\gamma^{d}. \tag{53}$$

This last statement is in agreement with (12). Note that

$$\sum_{r=1}^{N}\left(\mathbf{T}\mathbf{T}^{\dagger}\right)_{rr}=\int_{\mathbb{R}^3}d\mathbf{r}\ n_d(\mathbf{r})=\vartheta_x=\int_{\mathbb{R}^3}d\mathbf{r}\ n_a(\mathbf{r})=\sum_{s=1}^{L-N}\left(\mathbf{T}^{\dagger}\mathbf{T}\right)_{ss}. \tag{54}$$

It follows that $\vartheta_x=1$.

3.2. Detachment/attachment density matrix eigenvectors

This paragraph aims at demonstrating the connection between the NTOs and detachment/attachment paradigms by using the structure of the difference density matrix.

Theorem III.1 *NTOs are the eigenvectors of the detachment/attachment density matrices.*

Proof. We know from Lemma III.1 that

$$\gamma^{\Delta}=-\mathbf{T}\mathbf{T}^{\dagger}\oplus\mathbf{T}^{\dagger}\mathbf{T}=\overbrace{\begin{pmatrix}0_o & 0_{o\times v}\\ 0_{v\times o} & \mathbf{T}^{\dagger}\mathbf{T}\end{pmatrix}}^{\gamma^{a}}-\underbrace{\begin{pmatrix}\mathbf{T}\mathbf{T}^{\dagger} & 0_{o\times v}\\ 0_{v\times o} & 0_v\end{pmatrix}}_{\gamma^{d}};\quad \exists\mathbf{M}\mid\mathbf{M}^{\dagger}\gamma^{\Delta}\mathbf{M}=\mathbf{m}. \tag{55}$$

Since $\mathbf{TT}^\dagger$ and $\mathbf{T}^\dagger\mathbf{T}$ are positive definite, we deduce that the only negative eigenvalues of γ^Δ belong to the occupied × occupied block, while the positive ones belong to the virtual × virtual block. Since we know how to obtain the eigenvalues of $\mathbf{TT}^\dagger$ and $\mathbf{T}^\dagger\mathbf{T}$ thanks to Eq. (28), we know that the matrix $\mathbf{M}$ diagonalizing the difference density matrix must be the direct sum of $\mathbf{O}$ and $\mathbf{V}$:

$$\mathbf{m} = -\boldsymbol{\lambda}_o^2 \oplus \boldsymbol{\lambda}_v^2 \quad ; \quad \mathbf{M} = \mathbf{O} \oplus \mathbf{V}. \quad \blacksquare \tag{56}$$

According to Eq. (52), we deduce that the eigenvectors of the detachment/attachment density matrices are nothing but the occupied/virtual natural transition orbitals: the $\mathbf{M}^{d,a}$ matrices diagonalizing $\gamma^{d,a}$ are

$$\mathbf{M}^d = \mathbf{O} \oplus 0_v \;\Rightarrow\; \mathbf{M}^{d\dagger}\gamma^d\mathbf{M}^d = \boldsymbol{\lambda}_0^2 \oplus 0_v; \quad \mathbf{M}^a = 0_o \oplus \mathbf{V} \;\Rightarrow\; \mathbf{M}^{a\dagger}\gamma^a\mathbf{M}^a = 0_o \oplus \boldsymbol{\lambda}_v^2. \tag{57}$$

3.3. Equivalence of the two paradigms through quantitative analysis

Finally, and since we demonstrated that there is a direct relationship between the NTOs and the detachment/attachment, we will use Lemma III.1 and Theorem III.1 to demonstrate that our quantitative analysis is equivalent when derived in the two paradigms when the ground state wave function is a single Slater determinant and the excited state is a normalized linear combination of singly excited Slater determinants.

Corollary III.1 *The quantum descriptors derived from* γ^Δ *can be derived from* $\mathbf{T}$'s eigenvectors and singular values.

Proof. From Lemma III.1 and Theorem III.1, we can construct the following scheme:

$$\mathbf{O}^\dagger\mathbf{T}\mathbf{V} = \boldsymbol{\lambda} \rightarrow \{\boldsymbol{\lambda}_o^2; \boldsymbol{\lambda}_v^2\} \rightarrow \gamma^\Delta = -\mathbf{O}\boldsymbol{\lambda}_o^2\mathbf{O}^\dagger \oplus \mathbf{V}\boldsymbol{\lambda}_v^2\mathbf{V}^\dagger. \tag{58}$$

Following the structure of $\mathbf{m}$ deduced in Theorem III.1, we simply find $\mathbf{k}_\pm$

$$\mathbf{m} = -\boldsymbol{\lambda}_o^2 \oplus \boldsymbol{\lambda}_v^2 \Rightarrow \mathbf{k}_+ = 0_o \oplus \boldsymbol{\lambda}_v^2 \quad ; \quad \mathbf{k}_- = \boldsymbol{\lambda}_o^2 \oplus 0_v. \tag{59}$$

Backtransformation and few manipulations lead to

$$(\mathbf{O} \oplus \mathbf{V})\mathbf{k}_\pm(\mathbf{O}^\dagger \oplus \mathbf{V}^\dagger) = \gamma^{d,a} \rightarrow \{\phi_S, \tilde{\varphi}, \psi\}. \quad \blacksquare \tag{60}$$

'The joint computation of the NTOs and detachment/attachment density matrices from a single SVD, as a preliminary to the quantum metrics assessment, can even be simplified as

$$\mathbf{O}^\dagger\mathbf{T}\mathbf{V} = \boldsymbol{\lambda} \rightarrow (\mathbf{O} \oplus \mathbf{V})\begin{Bmatrix} \boldsymbol{\lambda}\boldsymbol{\lambda}^\dagger & \oplus & 0_v \\ 0_o & \oplus & \boldsymbol{\lambda}^\dagger\boldsymbol{\lambda} \end{Bmatrix}(\mathbf{O}^\dagger \oplus \mathbf{V}^\dagger) = \begin{Bmatrix} \gamma^d \\ \gamma^a \end{Bmatrix} \rightarrow \{\phi_S, \tilde{\varphi}, \psi\}. \tag{61}$$

Note finally that from Eq. (52) we see that the computation of the detachment/attachment density matrices (hence, the assessment of the topological metrics) can be performed without requiring any matrix diagonalization.

4. Conclusion

We rigorously detailed the theoretical background related to two methods allowing one to straightforwardly visualize how the absorption or emission of a photon impacts the electronic distribution of any complex molecular system. Based on one of these two methods, we showed that quantitative insights can be easily reached. Subsequently, we bridged the formalism of our two qualitative strategies in the case of single-reference excited states methods solely involving singly excited Slater determinants. Finally, it was demonstrated that in these cases any of the two qualitative methods can be used as a basis for deriving equivalent quantitative results. The totality of the features exposed in this book chapter is currently coded in the Nancy-Ex 2.0 [49] software suite and will be revisited, together with new strategies, in the TÆLES software [50] to be published soon.

Acknowledgements

Prof. Xavier Assfeld and Drs. Antonio Monari, Mariachiara Pastore, Benjamin Lasorne, Matthieu Saubanère, and Felix Plasser are gratefully acknowledged for fruitful discussions on the topic. Profs. Istvan Mayer, Anatoliy Luzanov, Martin Head-Gordon, and Andreas Dreuw are also thanked for their extremely inspiring work.

A. Derivation of the phase for $l = 2$ and 3

Figures 7 and **8** illustrate the evaluation process for the phase of terms 2 and 3 of Wick's expansion of the excited state density matrix elements in Eq. (41).

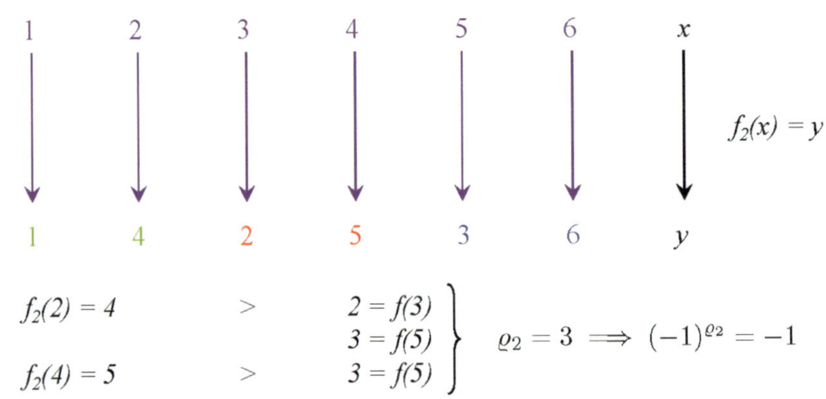

Figure 7. Illustration of the evaluation of ϱ_2.

http://dx.doi.org/10.5772/intechopen.70688

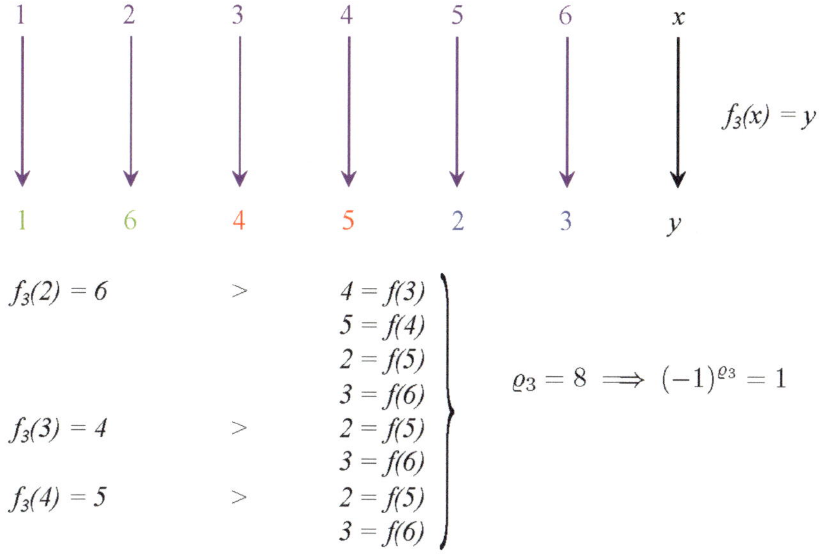

Figure 8. Illustration of the evaluation of ϱ_3.

B. Derivation of the equations in the atomic orbitals space

Most of the time, the spinorbitals themselves are expressed in a basis (often called basis of atomic orbitals, basis of atomic functions, or more simply a basis set) of K functions $\{\phi\}$. K might be superior to L when multiple spinorbitals in the atomic space are linearly dependent. The expression of spinorbitals in the atomic space is called linear combination of atomic orbitals (LCAO), and the pondering coefficients for a given spinorbital are stored in the column of a matrix, $\mathbf{C} \in \mathbb{R}^{K \times L}$, so that any spinorbital writes

$$\varphi_l(\mathbf{r}) = \sum_{\mu=1}^{K} (\mathbf{C})_{\mu l} \ \phi_\mu(\mathbf{r}). \tag{62}$$

Note that atomic orbitals are denoted by Greek letters for matrix elements. Since the spinorbitals correspond to columns in $\mathbf{C}$, we can split $\mathbf{C}$ into two matrices, $\tilde{\mathbf{O}} \in \mathbb{R}^{K \times N}$ and $\tilde{\mathbf{V}} \in \mathbb{R}^{K \times (L-N)}$, where the former contains the LCAO coefficients of the N first spinorbitals (the occupied ones) and the latter contains the LCAO coefficients for the last $L - N$ spinorbitals (the virtual ones). This splitting operation will be used later.

The spatial overlap between two atomic functions is also stored into a matrix, $\mathbf{S}$, which has the following elements:

$$(\mathbf{S})_{\mu\nu} = \int_{\mathbb{R}^3} d\mathbf{r} \ \phi_\mu^*(\mathbf{r})\phi_\nu(\mathbf{r}). \tag{63}$$

According to the LCAO expansion, the one-particle reduced density matrix kernel from Eq. (1) can be written in the atomic space for defining the density matrix $\mathbf{P}$ in the atomic space

$$\tilde{\gamma}(\mathbf{r}_1,\mathbf{r}_1') = \sum_{r=1}^{L}\sum_{s=1}^{L}\sum_{\mu=1}^{K}\sum_{\nu=1}^{K}\phi_\mu(\mathbf{r}_1)(\mathbf{C})_{\mu r}(\gamma)_{rs}(\mathbf{C})^*_{\nu s}\phi^*_\nu(\mathbf{r}_1') = \sum_{\mu=1}^{K}\sum_{\nu=1}^{K}\phi_\mu(\mathbf{r}_1)\underbrace{\sum_{r=1}^{L}\sum_{s=1}^{L}(\mathbf{C})_{\mu r}(\gamma)_{rs}(\mathbf{C}^\dagger)_{s\nu}}_{(\mathbf{P})_{\mu\nu}}\phi^*_\nu(\mathbf{r}_1')$$

$$= \sum_{\mu=1}^{K}\sum_{\nu=1}^{K}(\mathbf{P})_{\mu\nu}\phi_\mu(\mathbf{r}_1)\phi^*_\nu(\mathbf{r}_1'). \tag{64}$$

In these conditions, the number of electrons is given by the trace of **PS**. The central object for our investigations is now **P**, so that in the atomic space, the difference density matrix writes

$$\mathbf{\Delta} = \mathbf{P}^x - \mathbf{P}^0 \Rightarrow \mathrm{tr}(\mathbf{\Delta S}) = 0. \tag{65}$$

The difference density matrix in the atomic space can be diagonalized

$$\exists \mathbf{U} \mid \mathbf{U}^\dagger\mathbf{\Delta U} = \boldsymbol{\delta}. \tag{66}$$

Note here that $\boldsymbol{\delta}$ is a diagonal matrix containing the $\mathbf{\Delta}$ eigenvalues and should not be confused with the Kronecker delta. The $\mathbf{\Delta}$ eigenvalues can be sorted according to their sign:

$$\boldsymbol{\sigma}_\pm = \frac{1}{2}\left(\sqrt{\boldsymbol{\delta}^2} \pm \boldsymbol{\delta}\right) \tag{67}$$

and the resulting diagonal matrices can be separately backtransformed to provide the so-called detachment (**D**) and attachment (**A**) density matrices and the corresponding charge densities:

$$\mathbf{U}\boldsymbol{\sigma}_-\mathbf{U}^\dagger = \mathbf{D} \xrightarrow{\mathbb{R}^3} n_d(\mathbf{r}) = \sum_{\mu=1}^{K}\sum_{\nu=1}^{K}(\mathbf{D})_{\mu\nu}\ \phi_\mu(\mathbf{r})\phi^*_\nu(\mathbf{r}) \quad ; \quad \mathbf{U}\boldsymbol{\sigma}_+\mathbf{U}^\dagger = \mathbf{A} \xrightarrow{\mathbb{R}^3} n_a(\mathbf{r}) = \sum_{\mu=1}^{K}\sum_{\nu=1}^{K}(\mathbf{A})_{\mu\nu}\ \phi_\mu(\mathbf{r})\phi^*_\nu(\mathbf{r}). \tag{68}$$

From the detachment and attachment charge densities, one can then compute ϕ_S, $\tilde{\varphi}$, and ψ. Note that **D** and **A** should not be confused with "Donor" and "Acceptor" when dealing with push-pull dyes since, as we saw in **Figure 1**, the detachment or attachment densities are not strictly localized on fragments. Indeed, the detachment/attachment analysis is said to be systematic (or global), so is the quantitative analysis derived from it.

According to the structure of γ^Δ derived in Lemma III.1, and the connection between density matrices in the canonical and atomic spaces (see Eq. (64)), we can write $\mathbf{\Delta}$ using **T**:

$$\mathbf{\Delta} = \mathbf{C}\gamma^\Delta\mathbf{C}^\dagger = \mathbf{C}(-\mathbf{TT}^\dagger \oplus \mathbf{T}^\dagger\mathbf{T})\mathbf{C}^\dagger \tag{69}$$

which reduces to

$$\mathbf{\Delta} = \tilde{\mathbf{V}}(\mathbf{T}^\dagger\mathbf{T})\tilde{\mathbf{V}}^\dagger - \tilde{\mathbf{O}}(\mathbf{TT}^\dagger)\tilde{\mathbf{O}}^\dagger. \tag{70}$$

This means that from the transition density matrix one can easily reconstruct the difference density matrix in the atomic space, diagonalize it, and process until the obtention of the quantum metrics is achieved. This is the generalization of Corollary III.1 to the atomic space. We deduce from Eq. (69) that if K = L we have **U** = **SCM**.

Note finally that in the atomic space, occupied and virtual NTO LCAO coefficients are stored, respectively, in $\tilde{\mathbf{O}}\mathbf{O} \in \mathbb{R}^{K\times N}$ and $\tilde{\mathbf{V}}\mathbf{V} \in \mathbb{R}^{K\times(L-N)}$, where **O** and **V** are the left and right matrices implied in the SVD of **T** (see Eq. (24)).

Author details

Thibaud Etienne

Address all correspondence to: thibaud.etienne@umontpellier.fr

Institut Charles Gerhardt, CNRS and Université de Montpellier, Montpellier, France

References

[1] Plasser F, Lischka H. Analysis of excitonic and charge transfer interactions from quantum chemical calculations. Journal of Chemical Theory and Computation. Aug. 2012;**8**:2777-2789

[2] Guido CA, Cortona P, Adamo C. Effective electron displacements: A tool for time-dependent density functional theory computational spectroscopy. The Journal of Chemical Physics. Mar. 2014;**140**:104101

[3] Etienne T, Assfeld X, Monari A. Toward a quantitative assessment of electronic transitions charge-transfer character. Journal of Chemical Theory and Computation. Sept. 2014;**10**:3896-3905

[4] Etienne T, Assfeld X, Monari A. New insight into the topology of excited states through detachment/attachment density matrices-based centroids of charge. Journal of Chemical Theory and Computation. Sept. 2014;**10**:3906-3914

[5] Etienne T. Probing the locality of excited states with linear algebra. Journal of Chemical Theory and Computation. Feb 2015;**11**:1692-1699

[6] Guido CA, Cortona P, Mennucci B, Adamo C. On the Metric of Charge Transfer Molecular Excitations: A Simple Chemical Descriptor. Journal of Chemical Theory and Computation. July 2013;**9**:3118-3126

[7] Ciofini I, Le Bahers T, Adamo C, Odobel F, Jacquemin D. Through-Space Charge Transfer in Rod-Like Molecules: Lessons from Theory. The Journal of Physical Chemistry C. June 2012;**116**:11946-11955

[8] Ciofini I, Le Bahers T, Adamo C, Odobel F, Jacquemin D. Correction to "through-space charge transfer in rod-like molecules: lessons from theory". The Journal of Physical Chemistry C. July 2012;**116**:14736-14736

[9] Le Bahers T, Adamo C, Ciofini I. A Qualitative Index of Spatial Extent in Charge-Transfer Excitations. Journal of Chemical Theory and Computation. Aug. 2011;**7**:2498-2506

[10] García G, Adamo C, Ciofini I. Evaluating push–pull dye efficiency using TD-DFT and charge transfer indices. Physical Chemistry Chemical Physics. Oct. 2013;**15**:20210-20219

[11] Jacquemin D, Bahers TL, Adamo C, Ciofini I. What is the "best" atomic charge model to describe through-space charge-transfer excitations? Physical Chemistry Chemical Physics. Mar. 2012;**14**:5383-5388

[12] Cèron-Carrasco JP, Siard A, Jacquemin D. Spectral signatures of thieno [3, 4-b] pyrazines: Theoretical interpretations and design of improved structures. Dyes and Pigments. Dec. 2013;**99**:972-978

[13] Bäppler SA, Plasser F, Wormit M, Dreuw A. Exciton analysis of many-body wave functions: Bridging the gap between the quasiparticle and molecular orbital pictures. Physical Review A. Nov. 2014;**90**:052521

[14] Plasser F, Thomitzni B, Bäppler SA, Wenzel J, Rehn DR, Wormit M, Dreuw A, Statistical analysis of electronic excitation processes: Spatial location, compactness, charge transfer, and electron-hole correlation. Journal of Computational Chemistry. 2015;**36**(21):1620-1620

[15] Mewes SA, Plasser F, Dreuw A. Communication: Exciton analysis in time-dependent density functional theory: How functionals shape excited-state characters. The Journal of Chemical Physics. 2015;**143**(17):171101

[16] Etienne T, Assfeld X, Monari A. QM/MM modeling of Harmane cation fluorescence spectrum in water solution and interacting with DNA. Computational and Theoretical Chemistry. Feb. 2014;**1040-1041**:367-372

[17] Etienne T, Gattuso H, Michaux C, Monari A, Assfeld X, Perpète EA. Fluorene-imidazole dyes excited states from first-principles calculations–Topological insights. Theoretical Chemistry Accounts. 2016;**135**(4):1-11

[18] Etienne T, Chbibi L, Michaux C, Perpète EA, Assfeld X, Monari A. All-organic chromophores for dye-sensitized solar cells: A theoretical study on aggregation. Dyes and Pigments. Feb. 2014;**101**:203-211

[19] Duchanois T, Etienne T, Beley M, Assfeld X, Perpète EA, Monari A, Gros PC. Heteroleptic Pyridyl-Carbene Iron Complexes with Tuneable Electronic Properties. European Journal of Inorganic Chemistry. Aug. 2014;**2014**:3747-3753

[20] Dedeoglu B, Monari A, Etienne T, Aviyente V, Ozen AS. Detection of nitroaromatic explosives based on fluorescence quenching of silafluorene-and silole-containing polymers: A time-dependent density functional theory study. The Journal of Physical Chemistry C. Sept. 2014;**118**:23946-23953

[21] Ho EK-L, Etienne T, Lasorne B. Vibronic properties of para-polyphenylene ethynylenes: TD-DFT insights. The Journal of Chemical Physics. 2017;**146**(16):164303

[22] Ehrenreich H, Cohen MH. Self-consistent field approach to the many-electron problem. Physics Reviews. Aug 1959;**115**:786-790

[23] Hirata S, Head-Gordon M, Bartlett RJ. Configuration interaction singles, time-dependent Hartree–Fock, and time-dependent density functional theory for the electronic excited states of extended systems. The Journal of Chemical Physics. 1999;**111**(24):10774-10786

[24] Hirata S, Head-Gordon M. Time-dependent density functional theory within the Tamm–Dancoff approximation. Chemical Physics Letters. Dec. 1999;**314**:291-299

[25] Dreuw A, Head-Gordon M. Single-reference ab initio methods for the calculation of excited states of large molecules. Chemical Reviews. Nov. 2005;**105**:4009-4037

[26] Casida ME. In: Seminario JM, editor. Time-dependent density functional response theory of molecular systems: theory, computational methods, and functionals. Theoretical and Computational Chemistry, Vol. 4 of Recent Developments and Applications of Modern Density Functional Theory. Amsterdam, Elsevier; 1996. p. 391-439

[27] Casida ME. Time-dependent density-functional theory for molecules and molecular solids. Journal of Molecular Structure: THEOCHEM. Nov. 2009;**914**:3-18

[28] Head-Gordon M, Grana AM, Maurice D, White CA. Analysis of electronic transitions as the difference of electron attachment and detachment densities. The Journal of Physical Chemistry. 1995;**99**:14261

[29] Plasser F, Wormit M, Dreuw A. New tools for the systematic analysis and visualization of electronic excitations. I. Formalism. The Journal of Chemical Physics. July 2014;**141**:024106

[30] Plasser F, Bäppler SA, Wormit M, Dreuw A. New tools for the systematic analysis and visualization of electronic excitations. II. Applications. The Journal of Chemical Physics. July 2014;**141**:024107

[31] Pastore M, Assfeld X, Mosconi E, Monari A, Etienne T. Unveiling the nature of post-linear response Z-vector method for time-dependent density functional theory. The Journal of Chemical Physics. 2017;**147**(2):024108

[32] Etienne T. Transition matrices and orbitals from reduced density matrix theory. The Journal of Chemical Physics. 2015;**142**(24):244103

[33] Martin RL. Natural Transition Orbitals. The Journal of Chemical Physics. Mar. 2003;**118**: 4775-4777

[34] Batista, Martin, John Wiley & Sons. Encyclopedia of Computational Chemistry; 2004. https://goo.gl/P4qqt7

[35] Mayer I. Using singular value decomposition for a compact presentation and improved interpretation of the CIS wave functions. Chemical Physics Letters. Apr. 2007;**437**:284-286

[36] Surján PR. Natural orbitals in CIS and singular-value decomposition. Chemical Physics Letters. May 2007;**439**:393-394

[37] Furche F. On the density matrix based approach to time-dependent density functional response theory. The Journal of Chemical Physics. Apr. 2001;**114**:5982-5992

[38] Luzanov AV, Zhikol OA. Electron invariants and excited state structural analysis for electronic transitions within CIS, RPA, and TDDFT models. International Journal of Quantum Chemistry. 2010;**110**(4):902-924

[39] Luzanov AV, Sukhorukov AA, Umanskii V. Application of transition density matrix for analysis of excited states. Theoretical and Experimental Chemistry. July 1976;**10**:354-361

[40] Li Y, Ullrich CA. Time-dependent transition density matrix. Chemical Physics. 2001;**391**(1):157-163

[41] Tretiak S, Mukamel S. Density matrix analysis and simulation of electronic excitations in conjugated and aggregated molecules. Chemical Reviews. Sept. 2002;**102**:3171-3212

[42] Wu C, Malinin SV, Tretiak S, Chernyak VY. Exciton scattering approach for branched conjugated molecules and complexes. III. Applications. The Journal of Chemical Physics. 2008;**129**(17):174113

[43] Tretiak S, Igumenshchev K, Chernyak V. Exciton sizes of conducting polymers predicted by time-dependent density functional theory. Physical Review B. 2005;**71**(3):033201

[44] Amos AT, Hall GG. Single determinant wave functions. Proceedings of the Royal Society of London. Series A. Mathematical and Physical Sciences. Oct. 1961;**263**:483-493

[45] Frisch M et al. Gaussian 09 Revision B.01. Wallingford, CT: Gaussian Inc.; 2009

[46] Adamo C, Barone V. Toward reliable density functional methods without adjustable parameters: The PBE0 model. The Journal of Chemical Physics. Apr. 1999;**110**:6158-6170

[47] Adamo C, Scuseria GE, Barone V. Accurate excitation energies from time-dependent density functional theory: Assessing the PBE0 model. The Journal of Chemical Physics. Aug. 1999;**111**:2889-2899

[48] Frisch MJ, Pople JA, Binkley JS. Self-consistent molecular orbital methods 25. Supplementary functions for Gaussian basis sets. The Journal of Chemical Physics. Apr. 1984;**80**:3265-3269

[49] Assfeld X, Monari A, Very T, Etienne T. Nancy-Ex 2.0 software suite. http://nancyex.sourceforge.net

[50] Etienne T. TAELES software suite. http://taeles.wordpress.com

Chapter 4

Origin of Charge Transfer Exciton Dissociation in Organic Solar Cells

Shota Ono and Kaoru Ohno

Additional information is available at the end of the chapter

http://dx.doi.org/10.5772/intechopen.69854

Abstract

Using a temperature (T)-dependent tight-binding (TB) model for an electron-hole pair at the donor-acceptor (DA) interface, we investigate the dissociation of charge transfer exciton (CTE) into free carriers, that is, an electron and a hole. We observe the existence of the localization-delocalization transition at a critical T, below which the charges are localized to the DA interface, and above which the charges are delocalized over the system. This explains the CTE dissociation observed in organic solar cells. The present study highlights the combined effect of finite T and carrier delocalization in the CTE dissociation.

Keywords: charge transfer exciton, localization-delocalization transition, donor-accepter interface, tight-binding model, temperature

1. Introduction

Exciton, which is a two-particle state of electron and hole created by photon absorption of semiconductors or insulators, has been extensively studied since the seminal works of Frenkel [1, 2] and Wannier [3]. The binding energy of the exciton determines the photon absorption spectra near the band edges, where the Rydberg series, similar to the hydrogen-like excitation spectra, can be observed [4]. The concept of excitons is valid not only in solids but also in complex systems, such as nanostructures and interfaces. For example, let us consider two molecules with an appropriate separation. Given an electron-hole (EH) pair created in one molecule by a photon absorption, an electron in the molecule would be transferred to the other molecule due to the different lowest unoccupied molecular orbital (LUMO) energies, while a hole is left behind. Since the electron in the latter molecule and the hole in the former molecule interact with each other via the Coulomb interaction forces, they form a bound state, called as the charge transfer exciton (CTE) [5]. Recently, the CTE near the organic semiconductor interfaces has attracted much interest in the field of organic solar cells [6, 7]. This is a main concern in this chapter.

Organic solar cells, which generate electric power from the sunlight, play an important role in green energy industry and possess a variety of advantages: low cost, light, flexibility and easy-fabrication. The organic solar cells consist of the heterojunction between the electron donor and electron acceptor molecules. For example, in the C_{60}-based solar cells, the C_{60}-molecules serve as the accepter molecule and the organic thin-films such as X-phthalocyanine (XPc, X = Cu, Zn) [8–12] and single-walled carbon nanotubes [13, 14] serve as the donor molecules.

The principle of power generation in organic solar cells is decomposed into three steps, as shown in **Figure 1**: (i) exciton creation at the donor site by photon absorption, (ii) CTE creation following the movement of the created excitons to the donor-accepter (DA) interface and (iii) charge generation by the CTE dissociation into free carriers. While the second step may occur due to the different LUMO energies between the donor and acceptor molecules, the microscopic mechanism of the third step has not been understood; Since the CTE binding energy is a few hundreds of meV [6, 7, 15], the thermal energy is not enough to separate the EH pair into free carriers. In this way, several effects, such as dark dipoles [16, 17], disorder [18, 19], carrier delocalization [20–23], light effective mass [24] and entropy [25–28], on the CTE dissociation have been investigated. However, the relative importance of these factors is under debate.

In this chapter, we present an origin of the CTE dissociation by investigating the EH pair at the DA interface within a temperature (T)-dependent tight-binding (TB) model [29]. The important fact is that there exists a localization-delocalization transition at a critical T. The transition temperature estimated is in agreement with experimental observations in semiconductor interfaces [27]. Based on the T-dependence of the EH pair energy, we interpret the EH pair dynamics observed in time-resolved two-photon photoemission experiments [28]. Our model has shown that the transition can be observed only when the finite-T and the carrier delocalization effects are simultaneously considered. This review provides an important fact that more than one phenomenon might contribute to CTE dissociation.

The reminder of this chapter is organized as follows. In Section 2, we review the previous models of the CTE dissociation in organic solar cells. How the carrier delocalization effect is important in understanding the CTE dissociation is discussed. In Section 3, we present the formulation of the T-dependent TB model and the numerical results on the CTE dissociation. Our model is distinct from others in that the finite-T as well as the carrier delocalization effect is taken into account. In Section 4, how our model interprets the experimental data is discussed. Summary is presented in Section 5.

2. Literature review

We shall describe briefly some of the works that have theoretically discussed the origin of the CTE dissociation at the DA interface. The models can be classified into three levels on the basis of the approximation made (I) both charges, that is, electron and hole, are treated as localized particles [**Figure 2(a)**]; (II) one of the charges is treated as a delocalized particle, while the other is still treated as a localized one [**Figure 2(b)**] and (III) both charges are treated as delocalized

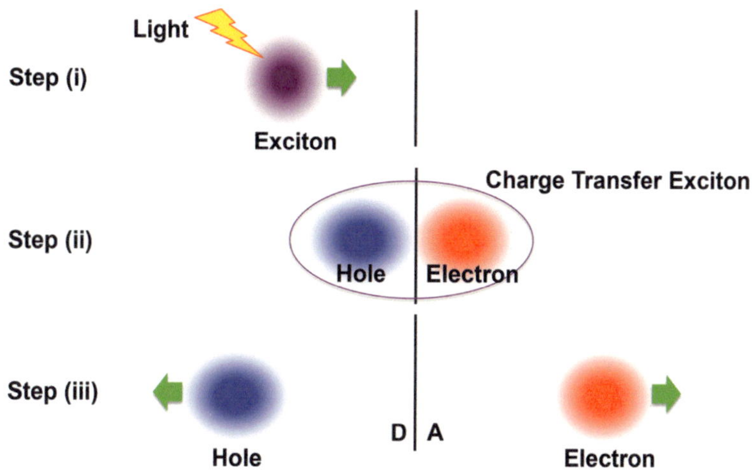

Figure 1. Schematic illustration of power generation in organic solar cells. The CTE, enclosed by an ellipse, consists of the electron and hole at the acceptor and donor, respectively. The donor and acceptor regions are abbreviated by D and A, respectively.

particles [**Figure 2(c)**], where the motion of the localized and delocalized particles would be described within the classical (or semi-classical) and quantum mechanics, respectively.

In the earliest study, Arkhipov et al. have constructed a dark dipole model within the approximation (I) above [16]. In this model, the DA interface consists of several polymer chains parallel to the DA interface. They computed the total energy of the CTE, that is, the sum of the electrostatic potential energy and the kinetic energy of the zero-point oscillations, by assuming the presence of the several dipoles at the DA interface. While the movement of the charged particle away from the interface lowers the Coulomb attractive forces, it also decreases the kinetic energy. They have found that the latter overcomes the former when the effective mass of the charged particle is less than $0.3m_e$, where m_e is the free electron mass, yielding the CTE dissociation. The effect of the different numbers of dipoles at the DA interface has also been investigated [17].

Deibel et al. have pointed out the importance of the charge delocalization along the polymer chains on the CTE dissociation by performing the kinetic Monte Carlo simulations [20]. To rationalize the concept of the delocalization, Nenashev et al. have developed an analytical model for the CTE dissociation within the approximation (II) [21]. They also studied the dissociation rate as a function of applied electric field by using the Miller-Abrahams expression for the hopping rate [30] and the dissociation probability formula for one-dimensional lattices [19]. The model has been further improved to include the effect of the dark dipoles at the DA interface [24]. However, those models have still employed the crude approximation (II) that one of the particles is fixed at a site.

Within the treatment (III), Raos et al. have computed the distribution of the electron and hole near the DA interface [22]. Using the TB approximation, they have shown that the sites where charge concentrates are not necessarily those just next to the DA interface, and this holds even

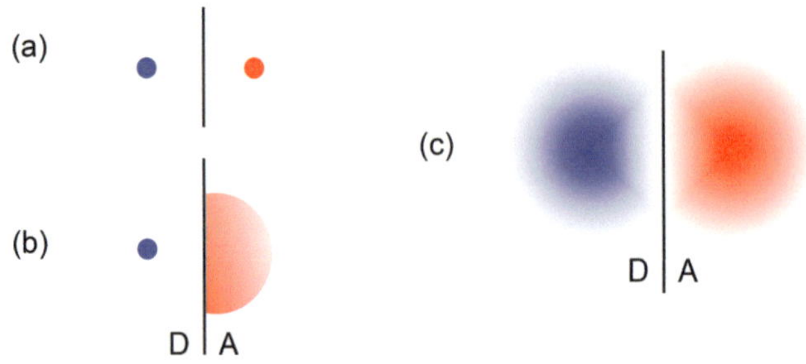

Figure 2. Schematic illustration of charged particles at the DA interface. (a) Localized hole and localized electron, (b) Localized hole and delocalized electron, and (c) delocalized hole and delocalized electron. Figures extracted and edited from Ref. [29].

in the ground state if diagonal and/or off-diagonal disorder exists. Athanasopoulos et al. have also confirmed that the CTE can efficiently dissociate into free carriers by extending the Arkhipov-Nenashev model above [23]. Recently, the authors have developed a *T*-dependent TB model applicable to the EH pair motion at the DA interface [29]. It has been shown that there exists a localization-delocalization transition of the EH pair at a critical *T*, below which the charges are localized to the DA interface, and above which the charges are delocalized over the system. This will be demonstrated below.

3. Localization-delocalization transition of EH pair

3.1. Formulation

We briefly provide the *T*-dependent TB model for describing the EH pair distribution at the DA interface. The details of the model have been provided in Ref. [29]. A similar approach has been used to study the size-dependent exciton energy of the quantum dots at zero *T* [31]. First, we consider an EH pair near the DA interface, assuming that only one photon is absorbed and that the electron-electron and hole-hole interaction energies are negligible. The electron and hole move around the acceptor and donor region, respectively, while they interact with each other via the attractive Coulomb interaction forces. Then, the Schrödinger equation for the two particles is given by

$$H^{(i)}|\phi_\alpha^{(i)}\rangle = \varepsilon_\alpha{}^{(i)}|\phi_\alpha^{(i)}\rangle, \tag{1}$$

where $\phi_\alpha^{(i)}$ and $\varepsilon_\alpha^{(i)}$ are the eigenfunction and eigenenergy with a quantum number α for the electron (i = e) and hole (i = h). Using the TB approximation, the Hamiltonian is given by

$$H^{(i)} = -\Sigma_{\boldsymbol{p},\boldsymbol{p}'} t_{\boldsymbol{p},\boldsymbol{p}'}^{(i)}|\boldsymbol{p}\rangle\langle\boldsymbol{p}'| + \Sigma_{\boldsymbol{p}} V_{\boldsymbol{p}}^{(i)}|\boldsymbol{p}\rangle\langle\boldsymbol{p}|, \tag{2}$$

where the first and second term denotes the kinetic and potential energies for the particle i, respectively. $t_{\boldsymbol{p},\boldsymbol{p}'}^{(i)}$ and $V_{\boldsymbol{p}}^{(i)}$ are the hopping integral between sites $\boldsymbol{p}$ and $\boldsymbol{p}'$ and the on-site potential energy at the site $\boldsymbol{p}$ = (p_x, p_y, p_z) with integers p_x, p_y, and p_z. The former is set to

$t^{(i)}_{p,p'} = t_0$, where t_0 is a positive constant, for simplicity. The effect of the long-range and anisotropic hopping has been investigated in Ref. [29]. The latter is explicitly given as

$$V_p^{(e)} = w_p^{(e)} - U_0 \Sigma_{p'} \frac{1}{|p - p'|} n_{p'}^{(h)}, \tag{3}$$

$$V_p^{(h)} = w_p^{(h)} - U_0 \Sigma_{p'} \frac{1}{|p - p'|} n_{p'}^{(e)}, \tag{4}$$

where U_0 determines the strength of the Coulomb interaction energy between the electron and hole. $w_p^{(i)}$ is the potential barrier height for the particle i, which will be given below. $n_p^{(i)}$ is the charge density for the particle i and is defined as

$$n_p^{(i)} = \Sigma_\alpha^{all} f_\alpha^{(i)} |\langle p | \phi_\alpha^{(i)} \rangle|^2, \tag{5}$$

where $|\langle p | \phi_\alpha^{(i)} \rangle|^2$ is the probability amplitude of the site p for the eigenstate $\phi_\alpha^{(i)}$. The summation is taken over the all eigenstates weighted by the Fermi distribution function $f_\alpha^{(i)}$ defined as

$$f_\alpha^{(i)} = \{\exp[\beta(\varepsilon_\alpha{}^{(i)} - \mu^{(i)})] + 1\}^{-1} \tag{6}$$

with the inverse temperature (β) and the chemical potential $\mu^{(i)}$, which will be determined by the relation of

$$\Sigma_p n_p^{(e)} = \Sigma_p n_p^{(h)} = 1. \tag{7}$$

The self-consistent solution of Eqs. (1)–(7) yields the electron and hole distributions near the DA interface. The solution enables us to compute the T-dependence of the free energy

$$\Omega = U_{\text{int}} - TS \tag{8}$$

with the internal energy

$$U_{\text{int}} = \Sigma_{i=e,h} \Sigma_\alpha \varepsilon_\alpha{}^{(i)} f_\alpha^{(i)} + U_0 \Sigma_p \Sigma_{p'} \frac{1}{|p - p'|} n_p^{(e)} n_{p'}^{(h)} \tag{9}$$

and the entropic energy

$$-TS = k_B T \Sigma_{i=e,h} \Sigma_\alpha \left[f_\alpha^{(i)} \ln f_\alpha^{(i)} + (1 - f_\alpha^{(i)}) \ln(1 - f_\alpha^{(i)}) \right], \tag{10}$$

where S denotes the entropy and k_B is the Boltzmann constant. Below, the hopping parameter t_0 will be used as an energy unit.

For later use, we define the charge density integrated over the p_x - p_y plane parallel to the interface

$$Q_{\text{tot}}^{(i)}(p_z) = \Sigma_{p_x} \Sigma_{p_y} n_p^{(i)} \tag{11}$$

with $i = e$ and h.

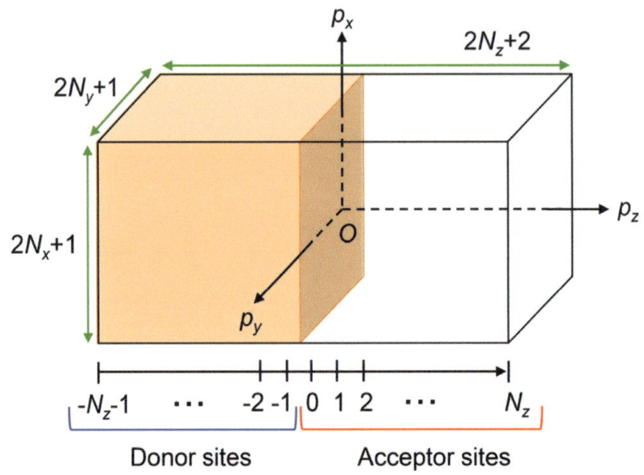

Figure 3. Simple cubic lattice for the DA interface model. The donor and acceptor regions are $-N_z - 1 \le p_z \le -1$ and $0 \le p_z \le N_z$, respectively. The total number of sites is $(2N_x + 1)(2N_y + 1)(2N_z + 2)$. Figure extracted from Ref. [29].

Figure 3 shows the DA interface model, where the simple cubic lattice is assumed. The movement of the electron and hole is restricted to the region of $-N_x \le p_x \le N_x$, $-N_y \le p_y \le N_y$, and $-N_z - 1 \le p_z \le N_z$. The potential barrier is assumed to be

$$w_{\boldsymbol{p}}^{(e)} = w_0 \theta(-0.5 - p_z), \tag{12}$$

$$w_{\boldsymbol{p}}^{(h)} = w_0 \theta(0.5 + p_z), \tag{13}$$

where $\theta(x)$ is the Heaviside step function, where $\theta(x) = 1$ for $x > 0$ and $\theta(x) = 0$ for $x < 0$. The numerical parameters in the model are set to $(N_x, N_y, N_z) = (5, 5, 10)$, $w_0 = 10t_0$, and $U_0 = 10t_0$, yielding the electron and hole that localize only to the acceptor and donor region, respectively at $T = 0$.

3.2. Numerical Results

Figure 4(a) shows the p_z-dependence of $Q_{\text{tot}}^{(e)}(p_z)$ and $Q_{\text{tot}}^{(h)}(p_z)$ in Eq. (11) for $k_B T/t_0 = 0$, 0.3, and 0.5. At zero T, $Q_{\text{tot}}^{(e)}(p_z)$ ($Q_{\text{tot}}^{(h)}(p_z)$) has the maximum value of 0.8 at $p_z = 0$ ($p_z = -1$) and decays within a few positive (negative) p_zs. As T increases, the p_z-dependence of $Q_{\text{tot}}^{(e)}(p_z)$ and $Q_{\text{tot}}^{(h)}(p_z)$ changes dramatically at around $k_B T/t_0 \simeq 0.3$: The values of $Q_{\text{tot}}^{(e)}(p_z)$ and $Q_{\text{tot}}^{(h)}(p_z)$ have the maximum of 0.3 at the sites away from those just next to the interface, that is, $p_z = 1$ and $p_z = -2$, respectively, and are averaged out over all p_z, which clearly indicate the CTE dissociation.

The localization-delocalization transition observed in **Figure 4(a)** can be understood as the free-energy anomaly. **Figure 4(b)** shows Ω in Eq. (8) as a function of T. The anomaly in Ω is observed at a critical temperature $k_B T_c/t_0 \simeq 0.27$. Ω is almost independent of T below T_c, while Ω decreases monotonically with increasing T above T_c. **Figure 4(c)** shows the T-dependence of the internal energy U_{int} and the entropy $-TS$ defined as Eqs. (9) and (10), respectively. Similar

http://dx.doi.org/10.5772/intechopen.69854

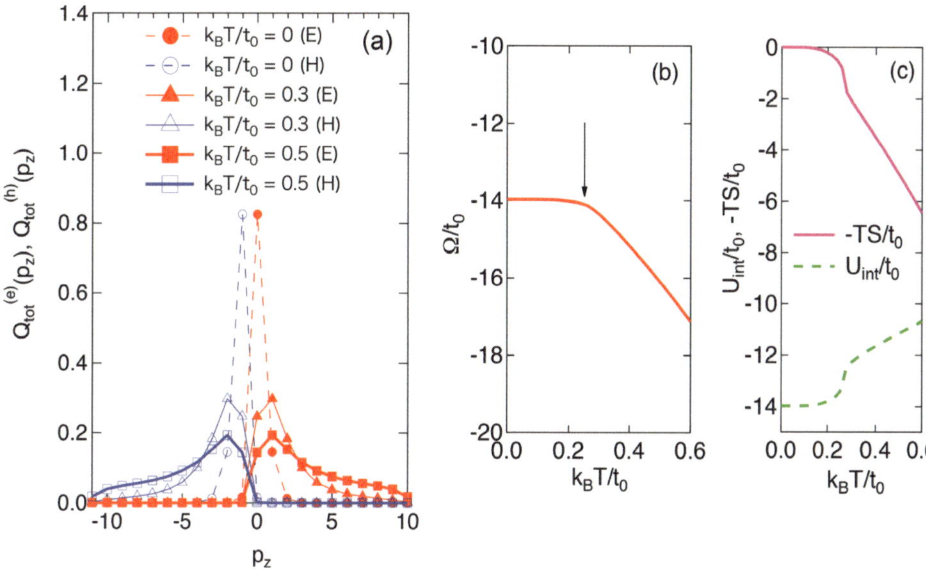

Figure 4. (a) The p_z dependence of $Q_{tot}^{(e)}$ (filled) and $Q_{tot}^{(h)}$ (open) given by Eq. (11) for $k_BT/t_0 = 0$ (circle), 0.3 (triangle), and 0.5 (square). The values of U_0/t_0 and w_0/t_0 are set to 10. (b) The T dependence of the free energy $\Omega(T)$ given by Eq. (8). The arrow indicates the free-energy anomaly that originates from the localization-delocalization transition. (c) U_{int} in Eq. (9) and $-TS$ in Eq. (10) as a function of T. Figures extracted and edited from Ref. [29].

anomalies are also observed in the T-dependence of U_{int} and $-TS$; U_{int} and $-TS$ jump at $T = T_c$, below which U_{int} and $-TS$ are almost independent of T, and above which U_{int} and $-TS$ increases and decreases, respectively. Since $S \simeq 0$ below T_c, Ω is dominated by the contribution from U_{int}. On the other hand, Ω is dominated by the entropy contribution above T_c.

To understand the microscopic mechanism of the localization-delocalization transition, we compute the density-of-states (DOS) for the EH pair, where the EH pair energy is defined as $E_\alpha{}^{(eh)} = \varepsilon_\alpha^{(e)} + \varepsilon_\alpha^{(h)}$. **Figure 5(a)** and **(b)** show the EH DOS at $k_BT/t_0 = 0$ and 0.6, respectively. At lower T, we can observe several peaks below the band edge: $E_\alpha{}^{(eh)} = -19.9, \ -17.1$ (doubly degenerate), and 16.4 eV. On the other hand, at higher T, no peaks are observed. **Figure 6(a)** shows the charge density of the electron and hole for the lowest 10 energy peaks at $T = 0$. The charge density is localized to the DA interface at lower $E_\alpha{}^{(eh)}$, while it is delocalized over the system at higher $E_\alpha{}^{(eh)}$. Note that at the lowest T the occupation probability of the lowest energy state is unity. When T is increased, the eigenvalue distribution changes. This is because the Fermi distribution function in Eq. (6) is broadened. This leads to the decrease in the Coulomb attractive forces between the electron and hole, yielding an upper shift of the EH pair energy. **Figure 6(b)** shows the T-dependence of $E_\alpha{}^{(eh)}$ for $\alpha = 1-10$. In fact, $E_\alpha{}^{(eh)}$ increases as T increases. The important fact is that the value of $E_\alpha{}^{(eh)}$ drastically increases at $T = 0.3t_0$, above which the energy level spacing is small compared to that below T_c. This yields the absence of peaks in the DOS near the band edge, shown in **Figure 5(b)**. The absence of isolated peaks means that all eigenstates are delocalized, indicating the localization-delocalization transition at a critical T.

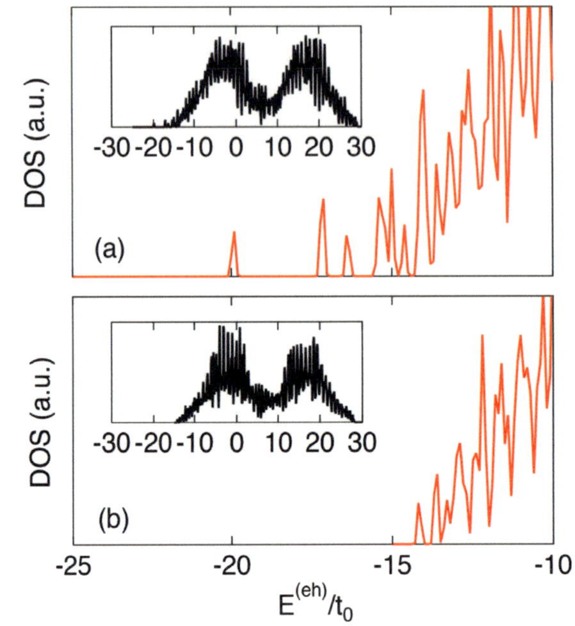

Figure 5. Electron-hole DOS (a) for $k_BT/t_0 = 0$ and (b) $k_BT/t_0 = 0.6$ from $E^{(eh)}/t_0 = -25$ to -10. The whole DOS is shown in the inset. Figures extracted from Ref. [29].

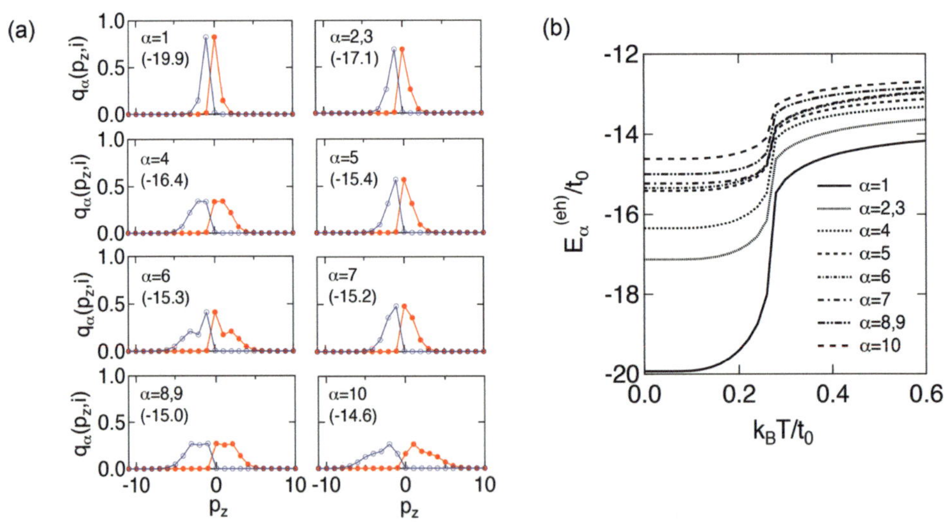

Figure 6. (a) The p_z dependence of electron (filled circle) and hole (open circle) density from the first to 10th eigenstate at $T = 0$. The eigenenergy $E_\alpha^{(eh)}$ is also shown in units of t_0. (b) $E_\alpha^{(eh)}$ as a function of T for $\alpha = 1$–10. The values of U_0/t_0 and w_0/t_0 are set to 10. Figures extracted and edited from Ref. [29].

The critical temperature increases significantly when one of the carriers is localized to only a site near the DA interface, that is, the approximation (II) is employed. This is because such a fixed charge enhances the attractive Coulomb interaction energy through Eqs. (3) and (4) and thus enhances the CTE binding energy significantly. The present result indicates that both the finite-T and the carrier delocalization effect are important to understand the CTE dissociation at the DA interface.

4. Discussion

4.1. Application to experiments

In this Section, we interpret the recent experimental observations on the CTE dissociation at the DA interfaces. Recently, Gao et al. have studied the charge generation in C_{60}-based organic solar cells through a measurement of the open-circuit voltage in a temperature range from 30 to 290 K. They have found that the number of free carriers created in the solar cells increases with increasing T, where the activation energy for the CTE dissociation is estimated to be 9 and 25 meV in annealed and unannealed systems, respectively [27]. To understand the magnitude of the activation energy, we compute the magnitude of T_c in typical organic solar cells. We set $U_0 = e^2/(4\pi\epsilon d) \simeq 0.5$ eV by using $\epsilon \simeq 3\epsilon_0$ (ϵ_0 is the dielectric constant of vacuum) and the equilibrium molecule-molecule distance $d \simeq 1$ nm of C_{60} crystals. The hopping parameter at the DA interface is set to $t_0 \simeq U_0/10$, by assuming that the single-particle band width is a few hundred meV. The height of the barrier potential is set to $w_0 \simeq U_0$, so that the CTE (not the Frenkel exciton) is formed at T = 0 K. Then, the value of $k_B T_c/t_0 \simeq 0.27$ corresponds to $k_B T_c \simeq 13$ meV. The magnitude of this energy is in agreement with the activation energy reported experimentally [27].

It is noteworthy that the magnitude of the CTE binding energy E_B can be estimated from the EH DOS. Assuming that the continuum states start from $\alpha \simeq 10$ at lower T shown in **Figure 5(a)**, the νth CTE binding energy is $E_B(\nu) = E^{(eh)}_{\alpha=10} - E^{(eh)}_{\alpha=\nu}$. For example, $E_B(v = 1) = 5.3\ t_0 \simeq 0.26$ eV, which is an order of magnitude higher than thermal energy at room temperature, but is consistent with experimental observations [15].

4.2. Scenario of the CTE dissociation

The agreement between our theory and experiments implies that the combined effect of the finite-T and carrier delocalization play a major role in the CTE dissociation. Based on our model, we show a possible scenario of the CTE dissociation in **Figure 7**.

1. The exciton is initially created at the donor region by photon absorption.
2. The electron transfer occurs at the DA interface, yielding the CTE formation.
3. The excess energy [32] created by the CTE formation (i.e. the energy difference between the donor LUMO and acceptor LUMO) excites phonons at the interface and disturbs the cold phonon distribution initially at T_0.
4. Through the phonon-phonon and phonon-electron scatterings, the phonon modes will obey the Bose distribution function with temperature T' higher than T_0 after the phonon thermalization time.
5. When T' is larger than T_c, the CTE can dissociate.

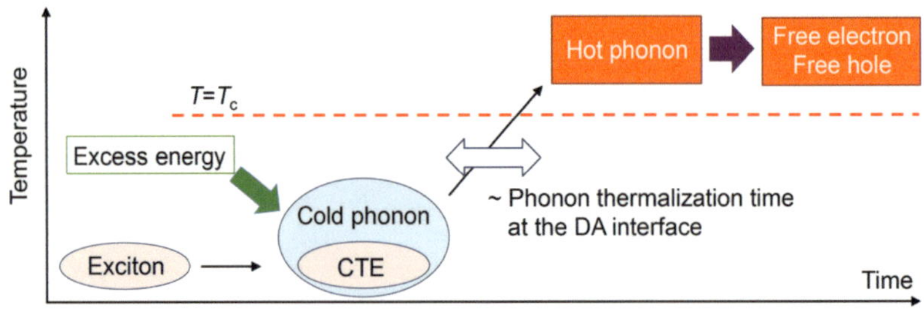

Figure 7. Temperature evolution of the charge carriers and phonons, provided that both the finite T mechanism proposed in the present study and the excess energy mechanism [32] hold. The excess energy created by the CTE formation excites the phonon at the DA interface. The phonon temperature increases with time and becomes over T_c within the phonon thermalization time. Above T_c, the CTE dissociates into the free electron and hole. Figure extracted from Ref. [29].

If this scenario holds, the magnitude of T' gradually increases with time. Then, the CTE energy also increases with time, as expected by the T-dependent $E_{\alpha}^{(eh)}$ shown in **Figure 6(b)**. This behaviour is quite similar to the experimental observations, where the CTE spontaneously climbs up the Coulomb potential at the pentacene-vacuum interface, by the time-resolved two-photon photoemission spectroscopy [28]. Such a CTE evolution has occurred within 100 fs that may be an order of the period of the optical phonon oscillations. For deeper understanding, it is necessarily to study the time-dependence of the interface phonon temperature T'. This may be studied in the framework of the non-equilibrium theory of phonons [33–35].

4.3. Some remarks

We also emphasize the finite-T effect on the excitonic properties. The exciton is usually described within many-body perturbation theory or time-dependent density-functional theory [36]. Recently, the CTE has been studied in such a first-principles context [37, 38]. The extension to the T-dependent Bethe-Salpeter or time-dependent Kohn-Sham equations and their solutions would give an accurate estimation of the CTE binding energy and predict the localization-delocalization transition or the free-energy anomaly mentioned in the present work.

In the present study, we have assumed that the dielectric constant is homogeneous across the DA interface. Recently, we have studied the effect of the inhomogeneity of the dielectric constant on the charge transfer behaviour within the continuum approach [39]. In such a system, the Coulomb interaction energy between two particles is given by

$$\frac{q_1 q_2}{4\pi\epsilon_0 \sqrt{\epsilon(\boldsymbol{r}_1)\epsilon(\boldsymbol{r}_2)}}, \tag{14}$$

where q_i and $\boldsymbol{r}_i$ are the charge and the position of the particle i. $\epsilon(\boldsymbol{r})$ is the local dielectric constant that describes the morphology of the DA interface. By solving the two-particle Schrödinger equation, we have demonstrated that the inhomogeneity of the dielectric constant yields an anisotropy of the charge distribution at the DA interface. Furthermore, we have found that the anisotropic distribution of the hole along the normal to the DA interface is important to yield the electron transfer, or vice versa. More investigation about the relation between the carrier distribution and the interface morphology is desired.

5. Summary

In this chapter, we have derived the T-dependent TB model for a EH pair at the DA interface, which enabled us to study the finite-T as well as the carrier delocalization effect on the CTE dissociation. Our numerical calculations have revealed that there exists the localization-delocalization transition at a critical temperature T_c, above which the CTE dissociates. This is related to the anomaly of the free energy Ω. Below and above T_c, Ω is determined by the internal energy and the entropic energy, respectively. The transition can be observed only when the carrier delocalization treatment is employed. The magnitude of T_c and the CTE binding energy estimated were in agreement with the experimental data. A possible scenario involving the phonon thermalization has been discussed.

So far, the origin of the CTE dissociation has been extensively investigated with consideration of a variety of models. Several effects on the CTE dissociation have been proposed, although the relative impact is not clear. The present study has emphasized the importance of the combined impact of the finite T and the carrier delocalization. Our work would be the first step for understanding the CTE dissociation observed at various DA interface in a unified manner. We hope that the localization-delocalization transition is observed in future experiments.

Acknowledgements

This work was supported by a Grant-in-Aid for Young Scientists B (No. 15K17435) from Japan Society for the Promotion of Science.

Author details

Shota Ono[1]* and Kaoru Ohno[2]

*Address all correspondence to: shota_o@gifu-u.ac.jp

1 Department of Electrical, Electronic and Computer Engineering, Gifu University, Gifu, Japan

2 Department of Physics, Graduate School of Engineering, Yokohama National University, Yokohama, Japan

References

[1] Frenkel J. On the transformation of light into heat in solids. I. Physical Review Journal. 1931;**37**:17

[2] Frenkel J. On the transformation of light into heat in solids. II. Physical Review Journal. 1931;**37**:1276

[3] Wannier G. The structure of electronic excitation levels in insulating crystals. Physical Review Journal. 1937;**52**:191

[4] Elliott RJ. Intensity of optical absorption by excitons. Physical Review Journal. 1957;**108**: 1384

[5] Halls JJM, Cornil J, Dos Santos DA, Silbey R, Hwang D-H, Holmes A B, Brédas JL, Friend RH. Charge- and energy-transfer processes at polymer/polymer interfaces: A joint experimental and theoretical study. Physical Review B. 1999;**60**:5721

[6] Few S, Frost JM, Nelson J. Models of charge pair generation in organic solar cells. Physical Chemistry Chemical Physics. 2015;**17**:2311

[7] Bässler H, Köhler A. "Hot or Cold": How do charge transfer states at the donor-acceptor interface of an organic solar cell dissociate? Physical Chemistry Chemical Physics. 2015;**17**:28451

[8] Hang H, Chen W, Chen S, Qi DC, Gao XY, Wee ATS. Molecular orientation of CuPc thin films on C_{60}/Ag(111). Applied Physics Letters. 2009;**94**:163304

[9] Shih C-F, Hung K-T, Chen H-J, Hsiao C-Y, Huang K-T, Chen S-H. Incorporation of potassium at CuPc/C_{60} interface for photovoltaic application. Applied Physics Letters. 2011;**98**:113307

[10] Lane PA, Cunningham PD, Melinger JS, Kushto GP, Esenturk O, Heilweil EJ. Photoexcitation dynamics in films of C_{60} and Zn phthalocyanine with a layered nanostructure. Physical Review Letters. 2012;**108**:077402

[11] Díaz AS, Burtone L, Riede M, Palomares E. Measurements of efficiency losses in blend and bilayer-type zinc phthalocyanine/C_{60} high-vacuum-processed organic solar cells. Journal of Physical Chemistry C. 2012;**116**:16384

[12] Lane PA, Cunningham PD, Melinger JS, Esenturk O, Heilweil EJ. Hot photocarrier dynamics in organic solar cells. Nature Communications. 2015;**6**:7558

[13] Dowgiallo A-M, Mistry KS, Johnson JC, Blackburn JL. Ultrafast spectroscopic signature of charge transfer between single-walled carbon nanotubes and C_{60}. ACS Nano. 2014;**8**:8573

[14] Ferguson AJ, et al. Trap-limited carrier recombination in single-walled carbon nanotube heterojunction with fullerene accepter layers. Physical Review B. 2015;**91**:245311

[15] Morteani AC, Sreearunothai P, Herz LM, Frend RH, Silva C. Exciton regeneration at polymeric semiconductor heterojunctions. Physical Review Letters. 2004;**92**:247402

[16] Arkhipov VI, Heremans P, Bässler H. Why is exciton dissociation so efficient at the interface between a conjugated polymer and an electron acceptor? Applied Physics Letters. 2003;**82**:4605

[17] Wiemer M, Nenashev AV, Jansson F, Baranovskii SD. On the efficiency of exciton dissociation at the interface between a conjugated polymer and an electron acceptor. Applied Physics Letters. 2011;**99**:013302

[18] Peumans P, Forrest SR. Separation of geminate charge-pairs at donor-acceptor interfaces in disordered solids. Chemical Physics Letters. 2004;**398**:27

[19] Rubel O, Baranovskii SD, Stolz W, Gebhard F. Exact solution for hopping dissociation of geminate electron-hole pairs in a disordered chain. Physical Review Letters. 2008; **100**:196602

[20] Deibel C, Strobel T, Dyakonov V. Origin of the efficient polaron-pair dissociation in polymer-fullerene blends. Physical Review Letters. 2009;**103**:036402

[21] Nenashev AV, Baranovskii SD, Wiemer M, Jansson F, Österbacka R, Dvurechenskii AV, Gebhard F. Theory of exciton dissociation at the interface between a conjugated polymer and an electron acceptor. Physical Review B. 2011;**84**:035210

[22] Raos G, Casalegno M, Idé J. An effective two-orbital quantum chemical model for organic photovoltaic materials. Journal of Chemical Theory and Computation. 2014;**10**:364

[23] Athanasopoulos S, Tscheuschner S, Bässler H, Köhler A. Efficient charge separation of cold charge-transfer states in organic solar cells through incoherent hopping. Journal of Physical Chemistry Letters. 2017;**8**:2093

[24] Schwarz C, Tscheuschner S, Frisch J, Winkler S, Koch N, Bässler H, Köhler A. Role of the effective mass and interfacial dipoles on exciton dissociation in organic donor-acceptor solar cells. Physical Review B. 2013;**87**:155205

[25] Clarke TM, Durrant JR. Charge photogeneration in organic solar cells. Chemical Reviews. 2010;**110**:6736

[26] Gregg B. A. Entropy of charge separation in organic photovoltaic cells: The benefit of higher dimensionality. Journal of Physical Chemistry Letters. 2011;**2**:3013

[27] Gao F, Tress W, Wang J, Inganäs O. Temperature dependence of charge carrier generation in organic photovoltaics. Physical Review Letters. 2015;**114**:128701

[28] Monahan NR, Williams KW, Kumar B, Nuckolls C, Zhu X-Y. Direct observation of entropy-driven electron-hole separation at an organic semiconductor interface. Physical Review Letters. 2015;**114**:247003

[29] Ono S, Ohno K. Combined impact of entropy and carrier delocalization on charge transfer exciton dissociation at the donor-acceptor interface. Physical Review B. 2016;**94**: 075305

[30] Miller A, Abrahams E. Impurity conduction at low concentrations. Physical Review. 1960; **120**:745

[31] Baskoutas S, Terzis AF, Schommers W. Size-dependent exciton energy of narrow band gap colloidal quantum dots in the finite depth square-well effective mass approximation. Journal of Computational and Theoretical Nanoscience. 2006;**3**:269

[32] Jackson NE, Savoie BM, Marks TJ, Chen LX, Ratner MA. The next breakthrough for organic photovoltaics? Journal of Physical Chemistry Letters. 2015;**6**:77

[33] Kabanov VV, Demsar J, Podobnik B, Mihailovic D. Quasiparticle relaxation dynamics in superconductors with different gap structures: Theory and experiments on $YBa_2Cu_3O_{7-\delta}$. Physical Review B. 1999;**59**:1497

[34] Ono S, H Shima, Y Toda. Theory of photoexcited carrier relaxation across the gap of phase-ordered materials. Physical Review B. 2012;**86**:104512

[35] Ono S. Nonequilibrium phonon dynamics beyond the quasiequilibrium approach. To appear in Physical Review B

[36] Onida G, Reining L, Rubio A. Electronic excitations: Density-functional versus many-body Green's-function approaches. Reviews of Modern Physics. 2002;**74**:601

[37] Cudazzo P, Sottile F, Rubio A, Gatti M. Exciton dispersion in molecular solids. Journal of Physics: Condensed Matter. 2015;**27**:113204

[38] Petrone A, Lingerfelt DB, Rega N, Li X. From charge-transfer to a charge-separated state: A perspective from the real-time TDDFT excitonic dynamics. Physical Chemistry Chemical Physics. 2014;**16**:24457

[39] Ono S, Ohno K. Minimal model for charge transfer excitons at the dielectric interface. Physical Review B. 2016;**93**:121301(R)

Chapter 5

Excitons and the Positronium Negative Ion: Comparison of Spectroscopic Properties

Sabyasachi Kar and Yew Kam Ho

Additional information is available at the end of the chapter

http://dx.doi.org/10.5772/intechopen.70474

Abstract

In view of the analogy of an exciton, biexciton and trion to the positronium (Ps) atom, Ps molecule, and Ps negative ion, in this chapter, we review our recent works on the Ps atom, Ps negative ion (Ps^-), and Ps-Ps interaction with Coulomb and screened Coulomb interactions for better understanding of spectroscopic properties of excitons, and excitonic ions and molecules. For the Coulomb case, this chapter describes the recent theoretical developments on the ground state, resonance states, photodetachment cross sections, polarizability and the recent experimental advancement on the efficient formation, photodetachment, resonance state of Ps^-. The chapter also presents results for the lowest $^3D^e$ Feshbach and $^1P^o$ shape resonances for Ps^- using correlated exponential wavefunctions. The $^1P^o$ shape resonance parameter is in agreement with the recent experiment. For screened interactions, various properties of Ps and Ps^- along with the dispersion coefficients for Ps-Ps interaction have been reviewed briefly. This review describes the effect of screened interactions on various properties of Ps^- within the framework of both screened Coulomb potential (SCP) and exponential-cosine-screened Coulomb potential (ECSCP). The influence of ECSCP on the dipole and quadrupole polarizability of Ps^- as functions of screening parameter and photon frequency are presented for the first time.

Keywords: excitons, positronium atom, trions, positronium negative ion, bi-excitons, positronium molecule, correlated exponential wave functions, spectroscopic properties, variational methods

1. Introduction

An exciton is a bound state of an electron and a positive hole (an empty electron state in a valence band), which is free to move through a nonmetallic crystal as unit. The electron and the positive hole are attracted to each other by the electrostatic Coulomb force. Excitons are electrically neutral quasiparticles that exist in insulators, semiconductors, and in some liquids. Excitons are difficult to detect as an exciton as a whole has no net electric charge, but the detection is possible by indirect means. Excitons can be described at various levels of sophistication; among them, the simplest and intuitive pictures can be understood using the effective mass approximation. Such approximation suggests that the Coulomb interaction between an electron and a positive hole leads to a hydrogen-like problem with a Coulomb potential term $-e^2/(4\pi\varepsilon_0\varepsilon|\mathrm{r}_e-\mathrm{r}_h|)$. Indeed, excitons in semiconductors form, to a good approximation, a hydrogen- or positronium-like series of states below the gap. The analogy of excitons to the hydrogen atom or even better the positronium atom can be pushed further. In analog to the formation hydrogen molecule or positronium molecule, two excitons can bind to form a new quasiparticle, the so-called bi-exciton or excitonic molecule. Similarly, in analog to the hydrogen molecular ion or the positronium negative ion, it is possible to form trions which are charged excitons or bi-excitons, i.e., quasiparticles of two electrons and one hole or vice versa. Like Ps molecule or Ps negative ion, bi-excitons or trions can also form bound states or quasi-bound states from the theoretical point of view. For detail discussions, classifications, and list of references on excitons, interested readers are referred to the review book authored by Klingshirn [1]. Keeping the above discussion in mind, it would be of great interest to review our works on the Ps atom, Ps negative ion, or Ps-Ps interaction for better understanding of spectroscopic properties of excitons, bi-excitons, or trions. The study of excitons under the influence of external environments is also of great interest both from theoretical and experimental sides. In this work, we have also discussed our recent study of the proposed systems under the influence screened Coulomb and cosine-screened Coulomb potentials.

The positronium negative ion (Ps^-) is the simplest bound three-lepton system (e^+, e^-, e^-) for which the $^1S^e$ state is the only state stable against dissociation but unstable against annihilation into photons. The Ps^- has gained increasing interest from the theoretical studies and experimental investigations since its theoretical prediction [2] and discovery [3]. This ion is a unique model system for studying three-body quantum mechanics as the three constituents of the Ps negative ion are subject only to the electroweak and gravitational forces. This elusive ion is of interest in the various branches of physics including solid-state physics, astrophysics, and physics of high-temperature plasmas, etc. It is also important for workability of many technical devices, such as modern communication devices. The Ps^- has been observed first by Mills [4] almost 40 years ago, and he subsequently measured its positron annihilation rate [5]. Since then, several experiments have been performed on this ion. Review of the most recent experiments can be found in the article of Nagashima [6] which also contains a large number of useful references. This review [6] also includes discussion on efficient formation of ion, its photodetachment, and the production of an energy-tunable Ps beam based on the technique of the photodetachment. It is here noteworthy to mention the accurate measurement of the decay rate [7] and only measurement of the $^1P^o$ shape resonance of Ps^- [8]. Several theoretical studies have been calculated so far on various properties of this ion, such as bound state [9–17], annihilation rate [16–18], photodetachment cross sections [19, 20], resonance states [21–24], and polarizability [25–27], using the numerical approaches such as the variational principle of Rayleigh-Ritz [9, 15–17, 28, 29], the correlation function hyperspherical harmonics method

[30–32], the complex-coordinate rotation method [33–36], the stabilization method [36–40], and the pseudostate summation method [25–27, 41–43]. Full list of articles can be found in the next sections. Besides such properties in the Coulomb case, several properties of the Ps negative ion have been studied under the influence of screened Coulomb potential (SCP) and exponential cosine-screened Coulomb potential (ECSCP). It is important to mention here that the study of atomic processes under the influence of screened interactions is an interesting, relevant, and hot topic of current research [44–49]. The complete SCP in a general form can be written as [50, 51]

$$V(r) = \begin{cases} Ze^2\left(\frac{1}{r} - \frac{1}{\lambda_D + \lambda_A}\right), & r \le \lambda_A \\ \left(\frac{\lambda_D}{\lambda_D + \lambda_A}\right)\frac{Ze^2}{r}\exp\left(-\frac{r-\lambda_A}{\lambda_D}\right), & r \ge \lambda_A \end{cases}, \tag{1}$$

where Z, λ_D, and λ_A denote the nuclear charge, the screening length, and the mean radius of the ion sphere, respectively. In the limit when $\lambda_A \to 0$, Eq. (1) reduces to the Debye-Hückel potential [52]. The ECSCP in form can be written as [53]

$$V(r) = \left(\frac{Ze^2}{r}\right)\exp(-\mu r)\cos(\mu r), \tag{2}$$

where μ is the screening parameter. The SCP or ECSCP occurs in several areas of physics (solid-state physics, ionized plasma, statistical thermodynamics, and nuclear physics). The potentials are also used in describing the potential between an ionized impurity and an electron in a metal or a semiconductor and the electron-positron interaction in a positronium atom in a solid [44–55]. In the next sections, we will briefly describe the properties of Ps negative ion, such as bound state, positron annihilation, resonance states, photodetachment, and polarizability. Bound states of the Ps atom and the Ps_2 molecule and dispersion coefficients on Ps-Ps interaction have also been discussed in the next sections.

2. Bound states

It is well-described that variational methods are the most effective and powerful tool for studying the Coulomb three-body bound-state problem [8, 11, 12, 16, 17, 56]. From here, we will concentrate on the works based on the variational approach. As mentioned in the last section, the Ps^- has very simple bound-state spectra that contain only one bound (ground), singlet state with total angular momentum, $L = 0$, i.e., $1\,^1S$ state for short. To calculate ground state energy of such ion, one needs to obtain the solutions of the Schrödinger equation, $H\Psi = E\Psi$, where $E<0$ following the Rayleigh-Ritz variational method. Here, we review our works using correlated exponential wave functions. The nonrelativistic screened Hamiltonian H (in atomic units) for a system having two electrons and a positron is given by

$$H = T + V, \tag{3}$$

with

$$T = -\frac{1}{2}\nabla_1^2 - \frac{1}{2}\nabla_2^2 - \frac{1}{2}\nabla_3^2, \tag{4}$$

$$V = -V(\mu, \mathbf{r}_{13}) - V(\mu, \mathbf{r}_{23}) + V(\mu, \mathbf{r}_{12}), \tag{5}$$

$$V(\mu, r_{ij}) = \frac{\exp\left(-\mu|\mathbf{r}_i - \mathbf{r}_j|\right)}{|\mathbf{r}_i - \mathbf{r}_j|} \cos\left(-\xi\mu|\mathbf{r}_i - \mathbf{r}_j|\right), \tag{6}$$

where 1 and 2 denote the two electrons and 3 denotes the positively charged particle and $|\mathbf{r}_i - \mathbf{r}_j| = r_{ji} = r_{ji} = |\mathbf{r}_j - \mathbf{r}_i|$. In Eq. (6), $\xi = 0$ for SCP, $\xi = 1$ for ECSCP, and $\mu = 0$ for unscreened case (UC). The variational wave functions for the ^{1}S-state of positronium negative ion can be shown as

$$\Psi_0(\mu) = \left(1 + \widehat{P}_{12}\right) \sum_{i=1}^{N_{B0}} C_i^0(\mu) \exp\left(-\alpha_i^0 r_{13} - \beta_i^0 r_{23} - \gamma_i^0 r_{12}\right), \tag{7}$$

where the operator $\widehat{P}_{12}$ is the permutation of the two identical particles 1 and 2. N_{B0} is the number of basis terms. The nonlinear variational parameters $\alpha_i^0, \beta_i^0, \gamma_i^0$ in the basis sets (7) are generated by the judicious implementation of a pseudorandom process of the following form

$$X_i^n = \left[\frac{1}{2} i(i+1)\sqrt{p_X}\right](R_{2,X} - R_{1,X}) + R_{1,X}, \tag{8}$$

$[x]$ is the fractional part of x, $[R_{1,X}, R_{2,X}]$ $(X = \alpha, \beta, \gamma)$ are real variational intervals which need to be optimized, and p_X assigns a separate prime number for each X. Quite a few theoretical studies have been performed to calculate binding energies of the proposed ion using variational wave functions (7) and the Hylleraas-type wave functions:

$$\Psi_{kmn} = \sum_{kmn} C_{kmn}\left(\exp[-\alpha(r_{13} + r_{23})] r_{12}^k r_{13}^m r_{23}^n + (1 \leftrightarrow 2)\right). \tag{9}$$

In Eq. (9), we also have $k + m + n \leq \Omega$, with Ω, l, m, and n being positive integers or zero. Detailed works in free atomic cases can be found from the earlier works [9–17, 57, 58]. In the screening environments, the ground state energy of Ps^- along with the electron affinity of Ps atom has been estimated variationally by Saha et al. [57] using multi-term correlated basis sets and SCP. The bound-state properties including ground state energies, radial and correlation cusp for this ion, and electron affinity of Ps have been investigated by us [58] using SCP and correlated wave functions (7). The bound states of Ps atom have also been described in our previous work under SCP ([59], references therein). To calculate the bound states of Ps atom, we have used standard Slater-type orbitals (see Eq. (40) in Section 7). Similar properties have been studied by Ghoshal and Ho [59] using ECSCP and wave function (9). The results show interesting behavior in the screening environments. The binding energies of the Ps molecule have been reported in previous works [60, 61].

3. Positron annihilation

The (e^+,e^-)-pair annihilation (or positron annihilation, for short) can proceed with the emission of a number of photons, for illustration, $e^+ + e^- = \gamma_1 + \gamma_2 + \gamma_3 \cdots + \gamma_K$, where γ_K is the emitted photons and K is the maximal number of such photons [16, 17]. Each of the annihilation processes has its unique annihilation width or annihilation rate $\Gamma_{k\gamma}$. For the proposed ion, the two-photon case would be the dominant annihilation process. However, the one-photon and three-photon, etc., annihilation are possible but in smaller rates. The annihilation rates $\Gamma_{2\gamma}$, $\Gamma_{3\gamma}$, $\Gamma_{4\gamma}$, $\Gamma_{5\gamma}$, and $\Gamma_{1\gamma}$ (arranged according to their numerical values) are important in applications. Here, we mention the formula for the one-, two-, three-, four-, and five-photon and total annihilation (Γ) rates, respectively [16, 17, 58]:

$$\Gamma_{1\gamma} = \frac{64\pi^2}{27}\alpha^8 c a_0^{-1} < \delta_{321} > = 1065.7569198 < \delta_{321} > s^{-1}, \tag{10}$$

$$\begin{aligned} \Gamma_{2\gamma} &= n\pi\alpha^4 c a_0^{-1}\left[1 - \frac{\alpha}{\pi}\left(5 - \frac{\pi^2}{4}\right)\right] < \delta(r_{31}) > \\ &= 100.3456053781 \times 10^9 < \delta(r_{31}) > s^{-1}, \end{aligned} \tag{11}$$

$$\begin{aligned} \Gamma_{3\gamma} &= n\alpha^5 c a_0^{-1}\frac{4(\pi^2 - 9)}{3} < \delta(r_{31}) > \\ &= 271.8545954 \times 10^6 < \delta(r_{31}) > s^{-1}, \end{aligned} \tag{12}$$

$$\Gamma_{4\gamma} \approx 0.274\left(\frac{\alpha}{\pi}\right)^2 \Gamma_{2\gamma}, \tag{13}$$

$$\Gamma_{5\gamma} \approx 0.177\left(\frac{\alpha}{\pi}\right)^2 \Gamma_{3\gamma}, \tag{14}$$

$$\begin{aligned} \Gamma \approx n(\Gamma_{2\gamma} + \Gamma_{3\gamma}) &= 2\pi\alpha^4 c a_0^{-1}\left[1 - \alpha\left(\frac{17}{\pi} - \frac{19\pi}{12}\right)\right] < \delta(r_{31}) > \\ &= 100.61745997357 \times 10^9 < \delta(r_{31}) > s^{-1}, \end{aligned} \tag{15}$$

where α, c, and a_0 denote, respectively, the fine structure constant, the velocity of light, and the Bohr radius and $<\delta_{321}>$ denotes the expectation value of three-particle delta function. It is obtained from the expectation value $<\Psi|\Psi>$ evaluated for $r_{32} = r_{31} = r_{21} = 0$. Exploiting the results for $<\delta_{321}>$ and $<\delta(r_{31})>$, one can easily calculate the values of $\Gamma_{1\gamma}$, $\Gamma_{2\gamma}$, $\Gamma_{3\gamma}$, $\Gamma_{4\gamma}$, $\Gamma_{5\gamma}$, and Γ using the explicit relation (10)–(15). The total annihilation rate along with the one-, two-, and three-photon annihilation rates, together with the values of $<\delta_{321}>$ and $<\delta(r_{31})>$ for various Debye lengths, is reported in our earlier work. The annihilation rates obtained from our calculations [59] are in agreement with the reported results [16, 17]. Detailed calculations of annihilation rate can be found from previous articles. As mentioned above, the positron annihilation process is of great interest in several areas of physics, such as astrophysics, solid-state physics, etc. It is also important for applicability of many technical devices, e.g., modern communication devices. In this review, we cited the recent references for free atomic case. For screened interaction, Kar and Ho [58] reported the annihilation rate under the influence of SCP, and Ghoshal and Ho [59] studied the similar features under ECSCP. The annihilation rates decrease with increasing screening strength.

4. Resonance states

A great number of theoretical studies on Ps^- have been performed in last few decades. Several studies have been performed on the resonances in e^--Ps scattering using the theoretical methods such as the Kohn-variational method [20], adiabatic treatment in the hyperspherical coordinates [62, 63], adiabatic molecular approximation [64], the hyperspherical close coupling method [65], the complex-coordinate rotation method [23, 24, 66–71], and the stabilization method [67, 68, 72–74]. For the recent advances in the theoretical studies on the resonances in Ps^-, readers are referred to recent reviews [23, 24, 66, 67, 75–77]. Review on resonance states of the proposed ion can be found in the articles of Ho [21–24, 33, 67–71]. Here, we review the resonance calculations using correlated exponential wave functions within the framework of two simple and powerful variational methods: the stabilization method (SM) and the complex-coordinate rotation method (CRM). The variational correlated exponential wave functions for higher partial wave states can be written as

$$\Psi_n(\mu) = \left(1 + S_{pn}\widehat{P}_{12}\right) \sum_{\substack{i=1 \\ l_1+l_2=L+\varepsilon}}^{N_{Bn}} \sum_{l_1=\varepsilon}^{L} C_i^n(\mu)(-1)^{\kappa} f(r_{13}, r_{23}, r_{21}) Y_{LM}^{l_1,l_2}(\mathbf{r}_{13}, \mathbf{r}_{12}), \tag{16}$$

with the radial function $f(r_{13}, r_{23}, r_{21})$ and the bipolar harmonics $Y_{LM}^{l_1,l_2}(\mathbf{r}_{13}, \mathbf{r}_{23})$,

$$f(r_{13}, r_{23}, r_{21}) = \exp\left[-\chi\left(\alpha_i^n r_{13} + \beta_i^n r_{23} + \gamma_i^n r_{21}\right)\right], \tag{17}$$

$$Y_{LM}^{l_1,l_2}(\mathbf{r}_{13}, \mathbf{r}_{23}) = r_{13}^{l_1} r_{23}^{l_2} \sum_{m_1, m_2} < l_1 l_2 m_1 m_2 | LM > Y_{l_1 m_1}(\widehat{r}_{13}) Y_{l_1 m_2}(\widehat{r}_{23}), \tag{18}$$

where $l_1 = i-(L+1)\text{mod}\{i/(L+1)\}$ for natural parity states, $l_1 = \text{mod}\{i/L\} + \kappa$ for unnatural parity states, mod{*i*/I} denotes the remainder of the integer division *i*/I, N_{Bn} is the number of basis terms, $\kappa = 0$ for natural parity states, $\kappa = 1$ for unnatural parity states, and χ is a scaling factor. Now, we would like to point out briefly the computational aspects of SM and CRM.

4.1. Computational aspect of SM

In the first step of resonance calculations using the stabilization method [37–40, 55, 67, 68, 72–74], it is mandatory to obtain precise values of energy levels. Resonance position can be

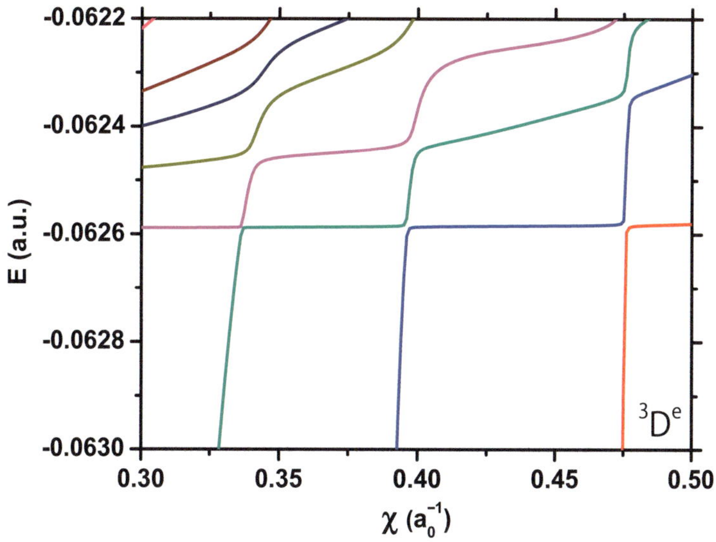

Figure 1. Stabilization diagram for the $^3D^e$ states of the Ps negative ion using 600 basis terms in Eq. (26).

identified after constructing stabilization diagram by plotting energy levels, E versus the scaling factor χ for certain μ. A stabilization diagram for the resonance states for the $^3D^e$ states of Ps^- for certain range of energy is depicted in **Figure 1**. The stabilized or slowly decreasing energy levels in the stabilization diagram indicate the position of the resonance at an energy E. Then to extract parameter (E_r, Γ) for a particular resonance state, one needs to calculate the density of the resonance states for each single energy level in the stabilization plateau using the formula

$$\rho_n(E) = \left| \frac{E_n(\alpha_{j+1}) - E_n(\alpha_{j-1})}{\alpha_{j+1} - \alpha_{j-1}} \right|^{-1}_{E_n(\alpha_j)=E}, \tag{19}$$

where the index j is the jth value for α and the index n is for the nth resonance. After calculating the density of resonance states $\rho_n(E)$ using formula (18), we fit it to the following Lorentzian form that yields resonance energy E_r and a total width Γ, with

$$\rho_n(E) = y_0 + \frac{A}{\pi} \frac{\Gamma/2}{(E - E_r)^2 + (\Gamma/2)^2}, \tag{20}$$

where y_0 is the baseline offset, A is the total area under the curve from the baseline, E_r is the center of the peak, and Γ denotes the full width of the peak of the curve at half height.

We obtained the desired results for a particular resonance state by observing the best fit (with the least chi-square and with the best value of the square of the correlation coefficient) to the Lorentzian form. The best fitting (solid line, using formula (20)) of the calculated density of states (*circles, using formula* (18)) for the lowest $^3D^e$ state of the Ps negative ion is presented in

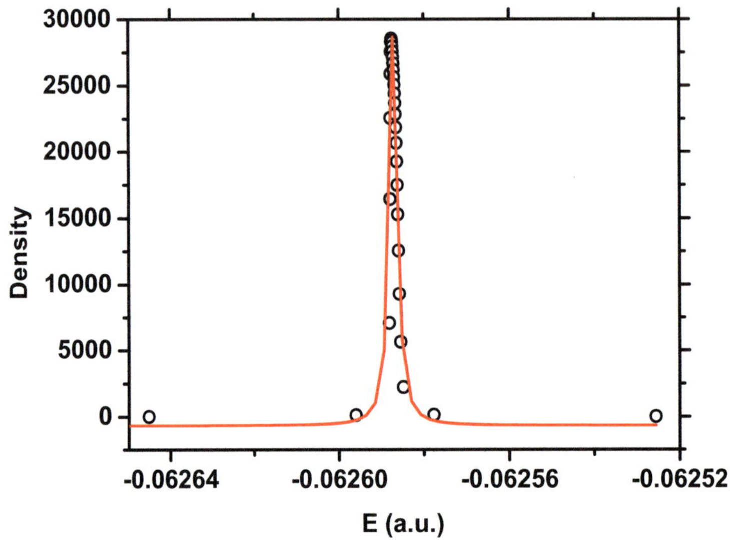

Figure 2. The best fitting (solid line) of the calculated density of states (*circles*) for the lowest $^3D^e$ state of the Ps negative ion.

Figure 2. The resonance position and width obtained from this work for the lowest $^3D^e$state below the Ps (N = 2) threshold as $E_r = -0.06259(1)$ a.u. and $\Gamma = 2.2(8) \times 10^{-6}$ a.u. are comparable with the results $E_r = -0.0625878(10)$ a.u. and $\Gamma = 6.4(20) \times 10^{-6}$ a.u. reported by Bhatia and Ho (see Refs. [70, 71]). As the $^3D^e$ resonance states are too narrow, so it seems difficult to extract resonance parameters for the other states above the Ps (N = 2) threshold. However, a $^3D^e$ resonance parameter is obtained for the first time using the stabilization method, as well as using correlated exponential wave functions.

4.2. Computational aspect of CRM

In the complex-rotation method [23, 24, 33], the radial coordinates are transformed by

$$r \to re^{i\theta} \tag{21}$$

and the transformed Hamiltonian takes the form:

$$H \to T\exp\left(-2i\theta\right) + Ve^{-i\theta}\exp\left(-re^{i\theta}\mu\right) \tag{22}$$

where T and V are the kinetic and the Coulomb part of potential energies. The wave functions are those of Eqs. (7) and (9). In the case of non-orthogonal functions, there are overlapping matrix elements:

$$N_{ij} = \left\langle \psi_i | \psi_j \right\rangle \tag{23}$$

and

$$H_{ij} = \left\langle \psi_i | H(\theta) | \psi_j \right\rangle \tag{24}$$

The complex eigenvalues problem can be solved with

$$\sum_i \sum_j C_{ij}(H_{ij} - EN_{ij}) = 0 \tag{25}$$

Resonance poles can be identified by observing the complex energy levels, $E(\theta,\alpha)$. The complex resonance eigenvalue is given by

$$E_{res} = E_r - \frac{i\Gamma}{2}, \tag{26}$$

where E_r is the resonance energy and Γ is the width. The resonance parameters are determined by locating stabilized roots with respect to the variation of the nonlinear parameters in the wave functions and of the rotational angle θ.

Resonance states for P, D, and F states of the Ps^- were reported following the abovementioned wave functions (16) and CRM [23, 24]. We have also located an S-wave shape resonances of the Ps^- lying above the Ps ($N = 2$) threshold using wave functions (18) and (9) and CRM [78]. Later, S-wave resonance states associated with and lying above the Ps ($N = 2, 3, 4, 5$) thresholds are reported by Jiao and Ho [79] using the wave function (9) and CRM. We have mentioned that a $^1P^o$ shape resonance has been observed in the laboratory [8]. The observed $^1P^o$ shape resonance is in agreement with the available theoretical data [80–82] and the present work using correlated exponential wave functions and CRM. **Figure 3** shows the rotational path for the $^1P^o$ shape resonance of the Ps^- lying above the Ps ($N = 2$) threshold, in the complex plane for four different values of the scaling factor, χ using 500-term correlated exponential basis functions. From this work, we have obtained the lowest $^1P^o$ shape resonance parameters as $E_r = -0.06212$

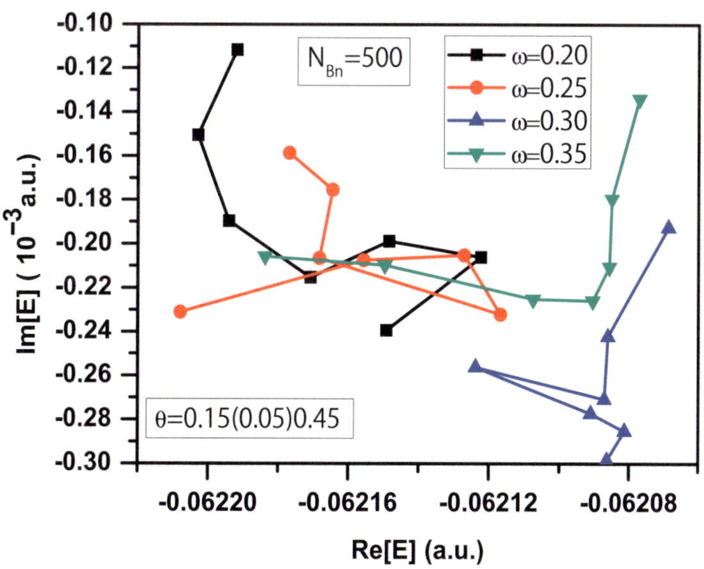

Figure 3. Rotational path of the $^1P^o$ shape resonance of the Ps^- lying above the Ps($N = 2$) threshold, in the complex plane for four different values of the scaling factor, χ using 500-term correlated exponential basis functions.

(3) a.u. and Γ = 0.00044(3) a.u. The numbers in the parentheses indicate the uncertainty in the last digits. The resonance states of Ps-Ps interaction were also studied by Ho [69].

In the screening environment, Kar and Ho [67, 68, 72–74] investigated the effects of SCP on the S-, P-, and D-wave resonance states of the Ps^- using correlated exponential wave functions, and Ghoshal and Ho [83] reported the effects of ECSCP on the lowest S-wave resonance state using the wave function (11) within the framework of SM. The resonance states have also successfully obtained using Hylleraas-type wave functions (9). Ho and Kar [76, 77] also investigated the S-wave resonance states of the proposed ion under the influence of SCP using CRM and wave function (9). In this work, wave functions (9) with up to $\Omega = 21$, $N_{B0} = 1078$, were used. The resonance parameters below the $N = 2, 3, 4, 5$, and 6 Ps thresholds, for various screening parameters, were reported. The lowest S-wave resonances of this ion interacting with ECSCP have also been studied by Ghoshal and Ho [83] using wave function (9) and ECSCP.

5. Photodetachment

The photoionization or photodetachment process is a subject of special interest in several areas of physics, such as astrophysics, plasma physics, and atomic physics due to its extreme importance in the atomic structures and correlation effects between atomic electrons [16, 17, 82, 84, 85]. The photoionization processes are also of great interest due to their applications in plasma diagnostics. Photodetachment of the Ps^- is also of particular interest as the experiments on Ps^- suggest that the Ps could be used to generate Ps beams of controlled energy, and this will involve acceleration of Ps^- and photodetachment of one electron. Photodetachment of the Ps^- is also of utmost importance due to its application in propagation of radiation in our galaxy. It is well known that the center of our galaxy, the Milky Way, contains a number of sources of the annihilation γ-quanta with $E_\gamma \approx 0.511$ MeV [86].

We reported the effect of screened Coulomb (Yukawa) potentials on the photodetachment cross sections of the positronium negative ion by using the asymptotic form of the bound-state wave function and a plane wave form for the final-state wave function. For detailed calculations and applications of the photodetachment of the positronium negative ion, interested readers are referred to the articles of Bhatia and Drachman [19], Frolov [17], Igarashi [82, 84, 85], Michishio et al. [8], Nagashima [6], and Ward et al. [20]. Here, we outlined the computational details in brief as mentioned in our earlier work [87] and in the works of Bhatia and Drachman [19].

In our previous work [87], we have considered the final-state wave function of the form $\Psi_f = \exp\left(i\,\vec{p}\,.\,\vec{r}\right)$ with $E = 3p^2/4$ and the initial bound-state wave function in the asymptotic region with the following form: $\Psi_i = C\exp(-\gamma r)/r$. The constant C for the Ps negative ion is obtained from the formula

$$C = G_A r \exp(\gamma r) \Psi_i(r, 0, r), \tag{27}$$

where G_A is some normalization constant and $\gamma = \sqrt{4(E_{Ps} - E_{Ps^-})/3}$, with E_{Ps^-} and E_{Ps}, the ground state energies of the Ps^- ion and Ps atom, respectively. The ground state energy of the Ps atom has been calculated using basis functions (40) prescribed in Section 7.

The photodetachment cross sections (σ) having photon energy E_p can be expressed as

$$\sigma = \frac{2}{3}\alpha a_0^2 p g(E_p) < \Psi_f|\widehat{\Lambda}(1,2)|\Psi_i > , \tag{28}$$

where α is the fine structure constant and $g(E) = E$ or E^{-1} for the dipole length and velocity approximations, respectively. The operator Λ represents the position and gradient operators for the length and velocity approximations, respectively, and can be written in explicit form as $\widehat{\Lambda}(1,2) = \Lambda\left(\vec{r}_{13}\right) + \Lambda\left(\vec{r}_{23}\right)$.

The final form of σ in terms of wavelength takes the form

$$\sigma = 4.30255225 \times 10^{-17} \rho^5 \frac{C^2}{\gamma^3}\left[\frac{\lambda}{\lambda_0}\right]^{3/2}\left[1 - \frac{\lambda}{\lambda_0}\right]^{3/2} cm^2, \text{with}\, \lambda \le \lambda_0, \tag{29}$$

and $\lambda_0 = 911.267057/\gamma^2$ (in Å), where ρ denotes the reduced electron mass. For the Ps^- ion, $\rho = \left(1 + M_c^{-1}\right)^{-1}$ with $M_c = 2$. The required normalization constant has been determined in this from highly accurate, completely non-adiabatic wave functions in Eq. (7) for the three-particle systems. Similar type of work was reported by Ghoshal and Ho using ECSCP and wave function (9) [88].

6. Polarizability

The study of atomic and ionic polarizabilities (both static and dynamic) plays an important role in a number of applications in physical sciences ([25–27, 44, 45, 89–98], references therein). When an atom or ion or molecule is placed in an electric field, the spatial distribution of its electrons experiences a distortion, the extent of which can be described in terms of its polarizability. The dynamic (dc) polarizability describes the distortion of the electronic charge distribution of an atom, ion, or molecule in the presence of an oscillating electric field of certain angular frequency. In this review, we describe the polarizability calculations of the Ps negative ion reported by Bhatia and Drachman [25], Kar and Ho [99], and Kar et al. [26, 27]. We also describe the polarizability calculations with SCP and ECSCP. To obtain dipole and quadrupole polarizability for the Ps^- ion, it is an important task to determine precisely the energies and wave functions for the ground state and the final P and D states. The dynamic 2^l-pole polarizability of the Ps^- ion in the screening environment can be written as [27]

$$\alpha_l(\omega) = \alpha_l^+(\omega) + \alpha_l^+(-\omega) \tag{30}$$

with

$$\alpha_l^+(\omega) = \frac{8\pi}{2l+1}\left(\frac{M}{M+1}\right)^{2l+1}\sum_n \frac{f_{nl}}{E_n(\mu) - E_0(\mu) + \omega} \text{ (in units of } a_0^{2l+1}), \tag{31}$$

where

$$f_{nl} = \left|\left\langle \Psi_0(\mu)\left|\sum_{i=1}^{2} r_i^l Y_{lm}(\mathbf{r}_i)\right|\Psi_n(\mu)\right\rangle\right|^2 \tag{32}$$

The summation in the above expression includes all the discrete and continuum eigenstates. Ψ_0 and Ψ_n describe the ground state eigenfunction with the corresponding energy eigenvalue E_0 and the nth intermediate eigenfunction for the final states with the corresponding eigenvalue, E_n, respectively. In the limit when $\omega \rightarrow 0$, $\alpha_l(\omega)$ is the static polarizability. For precise determination of eigenvalues and eigenfunction for each frequency and for each screening parameter for a particular system, one needs to solve the Schrödinger equation, $H\Psi = E\Psi$, by diagonalization of the Hamiltonian with the properly chosen wave functions in Eqs. (7) and (10). We rewrite the explicit form of wave function in Eq. (10) for polarizability calculations of this ion as

$$\Psi_n(\mu) = \left(1 + \widehat{P}_{12}\right) \sum_{\substack{i=1 \\ l_1+l_2=L}}^{N_{Bn}} C_i^n(\mu) \exp\left(-\alpha_i^n r_{13} - \beta_i^n r_{23} - \gamma_i^n r_{21}\right) Y_{LM}^{l_1,l_2}(\mathbf{r}_{13}, \mathbf{r}_{12}) \tag{33}$$

where $l_1 = i - (L+1)\text{mod}\{i/(L+1)\}$, $\text{mod}\{i/(L+1)\}$ denotes the remainder of the integer division $i/(L+1)$, and N_{Bn} is the number of basis term.

The static dipole and quadrupole polarizability for Ps^- has been reported by Bhatia and Drachman [25]. Kar and Ho also reported the static dipole polarizability of this ion in the screening environments as well in free atomic system [99]. Kar et al. also reported the dipole and quadrupole polarizabilities (static and dynamic) of this ion using SCP and exponential wave functions (33) [26, 27]. The dynamic dipole polarizability of the Ps^- was also studied by Kar et al. [27] in the screening environments. In this present work, we calculate the dipole and quadrupole polarizabilities (static and dynamic) under the influence of ECSCP and wave functions (33). The polarizabilities as functions of screening parameter and photon frequency are reported in **Figures 4** and **5** and **Tables 1** and **2**.

7. Dispersion coefficients for Ps-Ps interaction

Knowledge of the Van der Waals two-body dispersion coefficients in the multipole expansion of the second-order long-range interaction between a pair of atoms is of utmost importance for the quantitative interpretation of the equilibrium properties of gases and crystals, of transport phenomena in gases, and of phenomena occurring in slow atomic beams ([93, 100–102], references therein). The long-range part of the interaction potential between two spherically

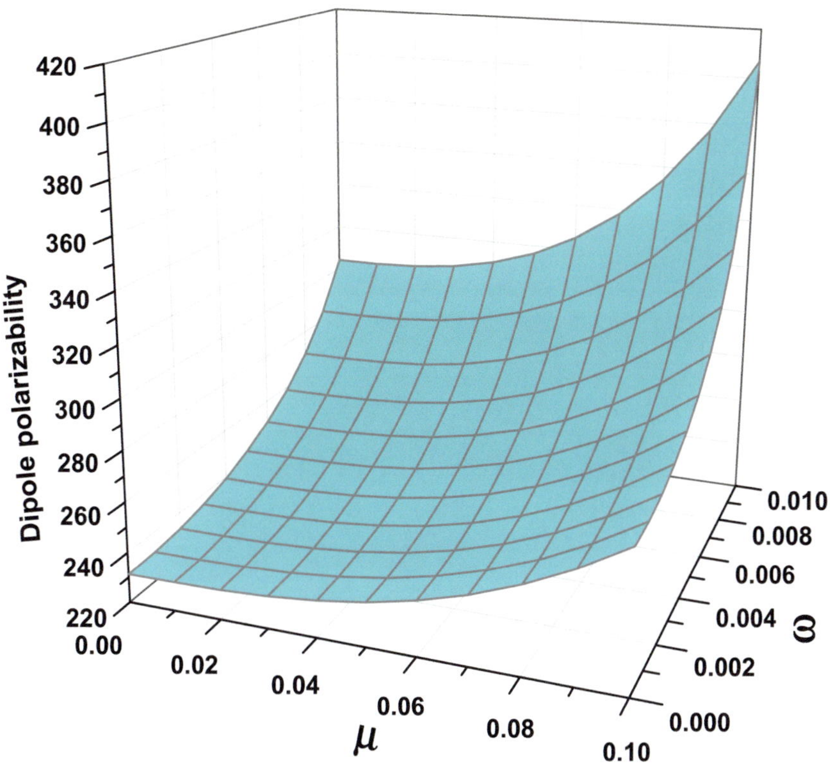

Figure 4. The dipole polarizability of the positronium negative ion as a function of screening parameter and photon frequency.

symmetric atoms a and b separated by a distance R can be written as a series with coefficients C_n denoted as dispersion coefficients [93, 100–102]:

$$V_{ab} = -\frac{C_6}{R^6} - \frac{C_8}{R^8} - \frac{C_{10}}{R^{10}} - \cdots, \tag{34}$$

with

$$C_6 = \frac{3}{\pi} G_{ab}(1,1), \tag{35}$$

$$C_8 = \frac{15}{2\pi}[G_{ab}(1,2) + G_{ab}(2,1)], \tag{36}$$

$$C_{10} = \frac{14}{\pi}[G_{ab}(1,3) + G_{ab}(3,1)] + \frac{35}{\pi} G_{ab}(2,2), \tag{37}$$

where

$$G_{ab}(l_a, l_b) = \frac{\pi}{2} \sum_{nm} \frac{f_{n0}^{(l_a)} f_{m0}^{(l_b)}}{E_{n0}^a E_{m0}^b \left(E_{n0}^a + E_{m0}^b\right)}, \tag{38}$$

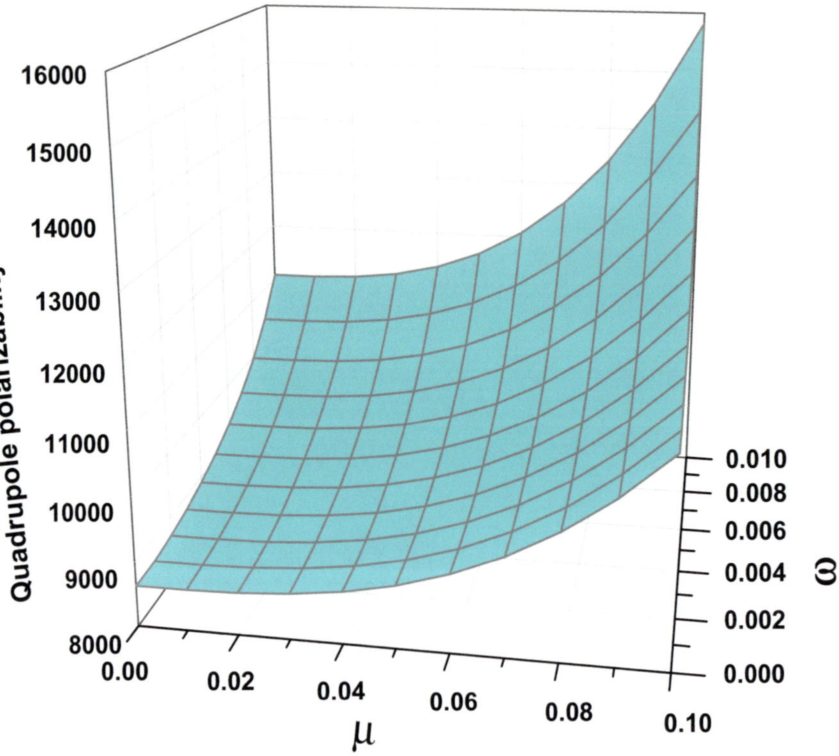

Figure 5. The quadrupole polarizability of the positronium negative ion as a function of screening parameter and photon frequency.

$E^i_{n0} = E^i_n - E^i_0$ is the excitation energy for atom i and is positive for the atoms in the ground state, and $f^{(l)}_{n0}$ denotes the 2^l-pole oscillator strengths and defined by

$$f^{(l)}_{n0} = \frac{8\pi}{2l+1}(E_n - E_0)\left|\left\langle < \Psi_0 \left| \sum_i r_i^l P_l(\cos\vartheta_i) \right| \Psi_n \right\rangle\right|^2, \tag{39}$$

with $i = 1$ for Ps and H atom. We also review here the dispersion coefficients for H-H interactions to establish a relation of dispersion coefficients with Ps-Ps and H-H interaction.

For positronium and hydrogen atoms, we have employed the Slater-type basis set:

$$\Psi = \frac{\sqrt{2l+1}}{4\pi}\sum_{i=l}^{N} D_i r^{i+l} e^{-\lambda r} P_l(\cos\theta_1), \tag{40}$$

where λ is the nonlinear variation parameters; l = 0, 1 for S and P states, respectively, and $D_i(i=1,\ldots.,N)$ are the linear expansion coefficients.

ω	μ = 0.01	μ = 0.02	μ = 0.04	μ = 0.05	μ = 0.06	μ = 0.08	μ = 0.09	μ = 0.10
0.000	231.3779	231.7355	234.4589	237.308	241.438	254.3094	263.4554	274.709
0.001	231.8534	232.2127	234.9495	237.813	241.964	254.9026	264.0985	275.416
0.002	233.2980	233.6626	236.4406	239.348	243.563	256.7081	266.0575	277.572
0.003	235.7687	236.1426	238.9923	241.976	246.302	259.8077	269.4255	281.285
0.004	239.3685	239.7562	242.7131	245.810	250.304	264.3508	274.3733	286.755
0.005	244.2600	244.6673	247.7751	251.033	255.763	270.5782	281.1787	294.311
0.006	250.6914	251.1256	254.4422	257.922	262.979	278.8678	290.2821	304.479
0.007	259.0429	259.5144	263.1203	266.908	272.423	289.8223	302.397	318.132
0.008	269.917	270.441	274.458	278.687	284.857	304.455	318.753	336.819
0.009	284.332	284.935	289.567	294.457	301.624	324.642	341.717	363.680
0.010	304.207	304.939	310.583	316.576	325.43	354.54	376.95	407.13
0.011	334.00	334.98	342.67	350.97	363.53	408.6	4.53[2]	
0.012	392.9	395.2	4.17[2]					

The numbers in square brackets indicate the power of 10.

Table 1. The dipole polarizability of the Ps negative ion for different screening parameters and photon frequencies.

ω	μ = 0.01	μ = 0.02	μ = 0.03	μ = 0.05	μ = 0.06	μ = 0.07	μ = 0.09	μ = 0.10
0.000	8630.1	8649.4	8701.3	8962.1	9198.5	9522.9	10496.4	11182.1
0.001	8647.3	8666.7	8718.8	8980.7	9218.0	9543.7	10521.4	11210.1
0.002	8699.5	8719.2	8771.9	9036.9	9277.2	9607.0	10597.3	11295.5
0.003	8788.6	8808.6	8862.4	9132.9	9378.2	9715.0	10727.4	11442.0
0.004	8917.6	8938.2	8993.7	9272.3	9525.0	9872.2	10917.3	1.1656[4]
0.005	9091.6	9113.1	9170.8	9460.7	9723.7	10085.4	1.1176[4]	1.1950[4]
0.006	9318.1	9340.7	9401.4	9706.6	9983.6	10364.9	1.1518[4]	1.2339[4]
0.007	9608.2	9632.3	9697.0	10022.7	1.0319[4]	1.0726[4]	1.1965[4]	12852[4]
0.008	0.9978[4]	1.0005[4]	1.0075[4]	1.0428[4]	1.0750[4]	1.1195[4]	1.2554[4]	1.3534[4]
0.009	1.0456[4]	1.0485[4]	1.0563[4]	1.0956[4]	1.1314[4]	1.1811[4]	1.3343[4]	1.4465[4]
0.010	1.1086[4]	1.1119[4]	11209[4]	1.1660[4]	1.2074[4]	1.2650[4]	1.446[4]	1.582[4]
0.011	1.196[4]	1.200[4]	1.211[4]	1.266[4]	1.317[4]	1.388[4]	1.62[4]	1.81[4]
0.012	1.330[4]	1.336[4]	1.350[4]	1.43[4]	1.50[4]	1.60[4]		

The numbers in square brackets indicate the power of 10.

Table 2. The quadrupole polarizability of the Ps^- in terms of screening parameter and photon frequency.

To investigate the effect on the dispersion coefficients C_6 in the screening environments, one can assume that the leading term in the Van der Waals interaction between two atoms a and b in their ground states still has a form of R^{-6}, as [101, 102]

$$V_{ab} = -\frac{C_6(\mu)}{R^6} + O(1/R^8) + \cdots. \tag{41}$$

Here, the plasma effect on V_{ab} is reflected on the value of C_6, which now depends on the screening parameter μ, and is denoted by $C_6(\mu)$. Similarly, to consider the plasma effect on the dispersion coefficients C_8 and C_{10}, we assume the coefficients depend on the screening parameter μ and are denoted, respectively, by $C_8(\mu)$ and $C_{10}(\mu)$. To calculate the dispersion coefficients for the interactions for Ps-Ps or H-H interactions, one needs to obtain the energy levels for the positronium atom or the hydrogen atom in the different partial wave states with the optimum choices of nonlinear parameters. To obtain the energy levels for hydrogen and positronium atoms with different Debye lengths, we diagonalize the Hamiltonian

$$H = -\frac{\eta}{2}\nabla^2 - \frac{\exp(-r/\lambda_D)}{r} \tag{42}$$

with the wave functions (40). Here, $\eta = 1$ is for the hydrogen atom and $\eta = 2$ for the positronium atom. In our previous work, we have reported the C_6, C_8, and C_{10} coefficients for Ps-Ps interactions under the influence of SCP. We have found from our calculations that the C_6, C_8, and C_{10} coefficients are, respectively, 2^5, 2^7, and 2^8 times larger than the corresponding coefficients of hydrogen-hydrogen interactions [103].

8. Comparison of spectroscopic properties and concluding remarks

To describe a semiconductor, one needs in principle to solve the Schrödinger equation for the problem. Depending on the coordinates of the ion cores having the nucleus and the tightly bound electrons in the inner shells and the outer or valence electrons with coordinates R_j and r_i and masses M_j and m_0, respectively, the Hamiltonian looks as ([1], Chapter 7)

$$H = -\frac{\hbar^2}{2}\sum_{j=1}^{M}\frac{1}{M_j}\nabla^2_{\mathrm{R}_j} - \frac{\hbar^2}{2m_0}\sum_{j=1}^{M}\nabla^2_{\mathrm{r}_i} + \frac{1}{4\pi\varepsilon_0}\left(\sum_{j>j'}\frac{e^2 Z_j Z_j}{\left|\mathrm{R}_j - \mathrm{R}_{j'}\right|} + \sum_{i>i'}\frac{e^2}{|r_i - r_{i'}|} + \sum_{i,j}\frac{e^2 Z_j}{|\mathrm{R}_j - r_i|}\right), \tag{43}$$

where Z_j is the effective charge of the ion core j and the indices j and i run over all M ion cores and N electrons, respectively. The wave function solving (43) can be constructed using all coordinates R_j and r_i including spins. The optical properties of the electronic system of a semiconductor or an insulator or even a metal can be understood as a description of the excited states of the N particle problem. The quanta of these excitations are known as "excitons" in semiconductors and insulators. The ground state of the electronic system for a perfect semiconductor can be described from various points of view as a completely filled valence band

and a completely empty conduction band [1]. However, from the theoretical side, the wave function of the bound state for excitons is said to be *hydrogenic*, an exotic atom (such as positronium atom) state akin to that of a hydrogen atom or even much better positronium atom. However, the binding energy is much smaller and the particle's size much larger than a hydrogen atom or larger than a positronium atom. This is due to the screening of the Coulomb force by other electrons in the semiconductor and due to the small effective masses of the excited electron and positive hole. However, it can be understood that the Hamiltonian for an exciton can be similar to a positronium atom if one can consider units using the Bohr radius for the respective system. The exciton Bohr radius is $a_B^{ex} = a_B^H \varepsilon \frac{m_0}{\tau}$ where the reduce exciton mass $\tau = \frac{m_e m_h}{m_e + m_h}$; m_e and m_h indicate the effective mass of electron and hole, respectively, and m_0 is the free electron mass. Exciton Rydberg energy is $\mathrm{Ry}^* = 13.6\mathrm{eV}\frac{\tau}{m_0}\frac{1}{\varepsilon^2}$ [1]. In similar way, the Hamiltonian for a trion and a bi-exciton can be related, respectively, with the Hamiltonian Ps negative ion and the Ps molecule. Wave functions for a trion or a bi-exciton could be similar with the Ps atoms or the Ps molecule. So, it is expected that the spectroscopic properties of the Ps atom, Ps negative ion, or Ps molecule might be useful to understand the spectroscopic properties of an exciton, trion, or bi-exciton.

Let us describe other types of comparison with bound excitons which are well studied in semiconductor, especially in gallium phosphide doped by nitrogen (GaP:N). The role and application of bound excitons in nanoscience and technology have been discussed in the article of Pyshkin and Ballato [104]. This investigation [104] observes something like neutral short-lived atom analog—a particle consisting of heavy negatively charged nucleus (N atom with captured electron) and a hole. Using bound excitons as short-lived analogs of atoms and sticking to some specific rules, Pyshkin and Ballato have been able to create a new solid-state media—consisting of short-lived nanoparticles excitonic crystal, obviously, with very useful and interesting properties for application in optoelectronics, nanoscience, and technology. Note that such specific rules include the necessity to build the excitonic superlattice with the identity period equal to the bound exciton Bohr dimension in the GaP:N single crystal. This study [104] also reports that the excitonic crystals yield novel and useful properties. These properties include enhanced stimulated emission and very bright and broadband luminescence at room temperature. With such development of bound excitons as short-lived analogs of atoms under some specific rules, it is also important to mention here that the emission spectra of representatives of exciton and positronium negative ion families can be realized from the earlier articles [104–108]. These articles support the usefulness of such comparisons of spectroscopic properties of excitons and the positronium negative ion. We hope that this chapter will provide a new direction and would be a remarkable reference for the future studies on excitons, bi-excitons, or trions as well as positronium, positronium molecule, and positronium negative ion.

Finally, we should also mention recent investigations on quantum information and quantum entanglement in few-body atomic systems, including the positronium negative ion. Quantification of Shannon information entropy, von Neumann entropy and its simpler form, linear entropy, for the two entangled (correlated) electrons in Ps^-, has been reported in the literature [109–111].

Acknowledgements

SK wishes to thank Prof. Z.C. Yan for his encouragement. SK also wishes to thank Ms. Yu-Shu Wang for her help, particularly in finding some references.

Author details

Sabyasachi Kar[1]* and Yew Kam Ho[2]

*Address all correspondence to: skar@hit.edu.cn

1 Department of Physics, Harbin Institute of Technology, Harbin, People's Republic of China

2 Institute of Atomic and Molecular Sciences, Academia Sinica, Taipei, Taiwan, Republic of China

References

[1] Klingshirn CF. Semiconductor Optics. 4th ed. Springer, Berlin, Heidelberg, Excitons, Biexcitons and Trions. Chapter 9. p. 249 and the chapters 7 (Crystals, Lattices, Lattice Vibrations and Phonons) and 8 (Electrons in a Periodic Crystal)

[2] Wheeler JA. Annals of the New York Academy of Sciences. 1946;**48**:221

[3] Hylleraas EA. Physical Review. 1947;**71**:491

[4] Mills AP Jr. Physical Review Letters. 1981;**46**:717

[5] Mills AP Jr. Physical Review Letters. 1983;**50**:671

[6] Nagashima Y. Physics Reports. 2014;**545**:95

[7] Cheeh H, Hugenschmidt C, Schreckenbach K, Gärtner SA, Thirolf PG, Fleischerand F, Schwalm D. Physical Review A. 2011;**84**:062508

[8] Michishio K, Kanai T, Kuma S, Azuma T, Wada K, Mochizuki I, Hyodo T, Yagishita A, Nagashima Y. Nature Communications. 2016;**7**:11060

[9] Hylleraas EA. Physical Review. 1946;**71**:491

[10] Barham M, Darewych JW. Journal of Physics B. 2008;**41**:185001

[11] Drake GWF, Grigorescu M. Journal of Physics B. 2005;**38**:3377

[12] Drake GWF, Cassar MM, Nistor RA. Physical Review A. 2002;**65**:054501

[13] Bhatia A, Drachman RJ. Nuclear Instruments and Methods in Physics Research Section B. 1998;**143**:195

[14] Frolov AM. Physical Review A. 1999;**60**:2834

[15] Ho YK. Physical Review A. 1993;**48**:4780

[16] Frolov AM. Physics Letters A. 2005;**342**:430

[17] Frolov AM. Chemical Physics Letters. 2015;**626**:49

[18] Puchalski M, Czarnecki A, Karshenboim SG. Physical Review Letters. 2007;**99**:203401

[19] Bhatia AK, Drachman RJ. Physical Review A. 1985;**32**:3745

[20] Ward SJ, Humberston JW, McDowell MRC. Journal of Physics B. 1987;**20**:127

[21] Ho YK. Resonances in positronium negative ions. In: A.K. Bhatia, editor. Proceedings for Temkin–Drachman Retirement Symposium, (NASA/CP-2006-214146), Goddard Space Flight Center, USA. 2006, p. 111

[22] Ho YK. Nuclear Instruments and Methods in Physics Research Section B: Beam Interactions with Materials and Atoms. 2008;**266**:516

[23] Kar S, Ho YK. The European Physical Journal D. 2010;**57**:13

[24] Kar S, Ho YK. Computer Physics Communications. 2011;**182**:119

[25] Bhatia AK, Drachman RJ. Physical Review A. 2007;**75**:062510

[26] Kar S, Li HW, Jiang P. Physics of Plasmas. 2013;**20**:083302

[27] Kar S, Wang Y-S, Wang Y, Ho YK. Polarizability of Negatively Charged Heliumlike Ions Interacting with Coulomb and Screened Coulomb potentials. International Journal of Quantum Chemistry. 2017

[28] Ritz W. Journal für die Reine und Angewandte Mathematik. 1909;**135**:1-61

[29] MacDonald JKL. Physical Review. 1933;**43**:830

[30] Mandelzweig VB. Nuclear Physics A. 1990;**508**:63C

[31] Bian W, Deng C. International Journal of Quantum Chemistry. 1994;**51**:285

[32] Krivec R, Mandelzweig VB, Varga K. Physical Review A. 2000;**61**:062503

[33] Ho YK. Physics Reports. 1983;**99**:1

[34] Reinherdt WP. Annual Review of Physical Chemistry. 1982;**33**:223

[35] Junker BR. Advances in Atomic and Molecular Physics. 1982;**18**:208

[36] Kar S, Ho YK. Physical Review A. 2007;**76**:032711

[37] Mandelshtam VA, Ravuri TR, Taylor HS. Physical Review Letters. 1993;**70**:1932

[38] Mandelshtam VA, Ravuri TR, Taylor HS. The Journal of Chemical Physics. 1994;**101**:8792

[39] Tan SS, Ho YK. Chinese Journal of Physics. 1997;**35**:701

[40] Kar S, Ho YK. Journal of Physics B. 2004;**37**:3177 and references therein

[41] Au CK, Drachman RJ. Physical Review A. 1988;**37**:1115

[42] Drake GWF. Physical Review Letters. 1970;**24**:765

[43] Victor GA, Dalgarno A, Taylor AJ. Journal of Physics B. 1968;**1**:13

[44] Kar S, Wang YS, Wang Y, Jiang Z. Few-Body Systems. 2017;**58**:14

[45] Kar S, Wang YS, Wang Y, Jiang Z. Physics of Plasmas. 2016;**23**:082119

[46] Ordóñez-Lasso AF, Martín F, Sanz-Vicario JL. Physical Review A. 2017;**95**:012504

[47] Ghoshal A, Ho YK. Physical Review A. 2017;**95**:052502

[48] Chaudhuri SK, Modesto-Costa L, Mukherjee PK. Physics of Plasmas. 2016;**23**:053305

[49] Janev RK, Zhang SB, Wang J. Matter and Radiation at Extremes. 2016;**1**:237

[50] Rouse CA. Physical Review. 1967;**163**:62

[51] Margenau H, Lewis M. Reviews of Modern Physics. 1959;**31**:569

[52] Debye P, Hückel E. Physikalische Zeitschrift. 1923;**24**:185

[53] Bessis N, Bessis G, Corbel G, Dakhel B. The Journal of Chemical Physics. 1975;**63**:3744

[54] Sil AN, Canuto S, Mukherjee PK. Advances in Quantum Chemistry. 2009;**58**:115

[55] Kar S, Ho YK. Journal of Physics B. 2009;**42**:044007 and references therein

[56] Korobov VI. Physical Review A. 2000;**61**:064503

[57] Saha B, Mukherjee TK, Mukherjee PK. Chemical Physics Letters. 2003;**373**:218

[58] Kar S, Ho YK. Chemical Physics Letters. 2006;**424**:403

[59] Ghoshal A, Ho YK. Few-Body Systems. 2009;**46**:249

[60] Hylleraas EA, Ore A. Physical Review. 1947;**15**:493

[61] Ho YK. Positronium Ions and Molecules. (NASA/N90-18990). USA: Goddard Space Flight Center; 1990. p. 243

[62] Botero J. Physical Review A. 1987;**35**:36

[63] Botero J. Zeitschrift für Physik D. 1988;**8**:177

[64] Rost JM, Wingten D. Physical Review Letters. 1992;**69**:2499

[65] Zhou Y, Lin CD. Physical Review Letters. 1995;**75**:2296

[66] Ho YK, Bhatia AK. Physical Review A. 1992;**45**:6268

[67] Kar S, Ho YK. Few-Body Systems. 2009;**45**:43

[68] Kar S, Ho YK. Few-Body Systems. 2009;**46**:173

[69] Ho YK. Hyperfine Interactions. 1992;**73**:109

[70] Ho YK. Chinese Journal of Physics. 1997;**35**:97

[71] Bhatia AK, Ho YK. Physical Review A. 1993;**48**:264

[72] Kar S, Ho YK. Physical Review A. 2005;**71**:052503

[73] Kar S, Ho YK. Physical Review A. 2006;**73**:032502

[74] Kar S, Ho YK. Nuclear Physics A. 2007;**790**:804c

[75] Ivanov IA, Ho YK. Physical Review A. 2000;**61**:032501

[76] Ho YK, Kar S. Few-Body Systems. 2012;**53**:437

[77] Ho YK, Kar S. Chinese Journal of Physics. 2016;**54**:574

[78] Kar S, Ho YK. Physical Review A. 2012;**86**:014501

[79] Jiao L, Ho YK. Few-Body Systems. 2013;**54**:1937

[80] Botero J, Greene CH. Physical Review Letters. 1986;**56**:1366-1369

[81] Ho YK, Bhatia AK. Physical Review A. 1993;**47**:1497-1499

[82] Igarashi A, Shimamura I, Toshima N. New Journal of Physics. 2000;**2**:17

[83] Ghoshal A, Ho YK. The European Physical Journal D. 2010;**56**:151

[84] Igarashi A. Few-Body Systems. 2017;**58**:1-5

[85] Igarashi A, Shimamura I. Journal of Physics B. 2004;**37**:4221

[86] Sivaram C, Krishan V. Astrophysics and Space Science. 1982;**85**:31

[87] Kar S, Ho YK. Few-Body Systems. 2008;**42**:73-81

[88] Ghoshal A, Ho YK. Few-Body Systems. 2010;**47**:185

[89] Mitroy J, Safronova MS, Clark CW. Journal of Physics B. 2010;**43**:202001

[90] Sen S, Mandal P, Mukherjee PK. Physics of Plasmas. 2012;**19**:033501

[91] Kar S. Physical Review A. 2012;**86**:062516

[92] Li HW, Kar S. Physics of Plasmas. 2012;**19**:073303

[93] Kar S, Li HW, Shen ZC. Journal of Quantitative Spectroscopy and Radiative Transfer. 2013;**116**:34

[94] Li HW, Kar S, Jiang P. International Journal of Quantum Chemistry. 2013;**113**:1493

[95] Zaleśny R, Goŕ A, Kozłowska J, Luis JM, Ågren H, Bartkowiak W. Journal of Chemical Theory and Computation. 2013;**9**:3463

[96] Kar S, Kamali MZM, Ratnavelu K. AIP Conference Proceedings. 2014;**1588**:87

[97] Kar S, Wang Y-S, Li W-Q, Sun X-D. International Journal of Quantum Chemistry. 2015;**115**:1573

[98] Henson BM, Khakimov RI, Dall RG, Baldwin KGH, Tang L-Y, Truscott AG. Physical Review Letters. 2015;**115**:043004

[99] Kar S, Ho YK. Physics Letters A. 2008;**372**:4253

[100] Yan Z-C, Babb JF, Dalgarno A, Drake GWF. Physical Review A. 1996;**54**:2824

[101] Kar S, Ho YK. Chemical Physics Letters. 2007;**449**:246

[102] Kar S, Ho YK. Physical Review A. 2010;**81**:062506

[103] Kar S, Ho YK. Nuclear Instruments and Methods in Physics Research, Section B. 2008;**266**:526

[104] Pyshkin SL, Ballato J. Optoelectronics: Advanced Devices Structures. InTech, Chapter 1; 2017

[105] Pyshkin SL. Optoelectronics: Materials and Devices. InTech, Chapter 1; 2015

[106] Pyshkin SL, Ballato J. Optoelectronics: Advanced Materials and Devices. InTech, Chapter 1; 2013

[107] Gaspari F. Optoelectronics: Material and Techniques. InTech, Chapter 1; 2015

[108] Krause-Rehberg R, Leipner HS. Positron Annihilation in Semiconductors: Defect Studies. Berlin: Springer-Verlag; 1998

[109] Lin C-H, Ho YK. Physics Letters A. 2014;**378**:2861-2865

[110] Lin C-H, Ho YK. Chemical Physics Letters. 2015;**633**:261-264

[111] Lin C-H, Ho YK. Atoms. 2015;**3**:422-432